TOWARDS A CLIMATE-NEUTRAL EUROPE

This book explains the EU's climate policies in an accessible way, to demonstrate the step-by-step approach that has been used to develop these policies and the ways in which they have been tested and further improved in the light of experience. The latest changes to the legislation are fully explained throughout.

The chapters throughout this volume show that no single policy instrument can bring down greenhouse gas emissions. The challenge facing the EU, as for many countries that have made pledges under the Paris Agreement, is to put together a toolbox of policy instruments that is coherent, delivers emissions reductions and is cost-effective. The book stands out because it covers the EU's emissions trading system, the energy sector and other economic sectors, including their development in the context of international climate policy.

This accessible book will be of great relevance to students, scholars and policy makers alike.

Jos Delbeke is a Professor at the School of Transnational Governance at the European University Institute in Florence and at the University of Leuven (KU Leuven), Belgium, from which he holds a PhD in economics (1986). He was Director-General of the European Commission's Directorate-General for Climate Action from its creation in 2010 to 2018, and he has worked for the European Commission since 1986.

Peter Vis retired from the European Commission in 2019. He was previously an Adviser in the Commission's in-house think-tank, the European Political Strategy Centre. Over his career he worked in several Directorates-General of the European Commission. For the academic year 2014–2015, he was the EU Visiting Fellow at St Antony's College, University of Oxford, UK. He was Chief of Staff to Connie Hedegaard, European Commissioner for Climate Action (2010–2014). He has an MA (History) from the University of Cambridge, UK.

TOWARDS A CLIMATE-NEUTRAL EUROPE

Curbing the Trend

Edited by Jos Delbeke and Peter Vis

LONDON AND NEW YORK

First published 2019
by Routledge
2 Park Square, Milton Park, Abingdon, Oxon OX14 4RN

and by Routledge
52 Vanderbilt Avenue, New York, NY 10017

Routledge is an imprint of the Taylor & Francis Group, an informa business

British Library Cataloguing-in-Publication Data
A catalogue record for this book is available from the British Library

Library of Congress Cataloging-in-Publication Data
A catalog record for this book has been requested

ISBN: 978-9-276-09260-5 (hbk)
ISBN: 978-9-276-08261-3 (pbk)
ISBN: 978-9-276-08256-9 (ebk)

Typeset in Bembo
by Apex CoVantage, LLC

MIX
Paper from
responsible sources
FSC
www.fsc.org   FSC™ C013985

Printed in the United Kingdom
by Henry Ling Limited

CONTENTS

3.2.2 Milestones *49*
3.2.3 Sectoral analysis *50*
3.2.4 Investments *53*
3.2.5 Innovation *54*
3.3 Towards a Commission proposal for a 2030 climate and energy framework *54*
   3.3.1 Green Paper consulting on the possible 2030 framework *55*
   3.3.2 The Impact Assessment *56*
   3.3.3 The 2014 Commission Communication on the 2030 framework *59*
   3.3.4 The European Council Conclusions of October 2014 *61*
Conclusion *64*

4 The EU Emissions Trading System   66
Damien Meadows, Peter Vis and Peter Zapfel

Introduction *66*
4.1 How does the EU Emissions Trading System work? *66*
4.2 Emission reductions of 26% under the EU ETS from 2005–2017 *68*
4.3 Addressing the low carbon price since 2013 *70*
4.4 A fundamental review for the period from 2021–2030 *73*
4.5 Carbon leakage and free allocation: having industry on board *75*
   4.5.1 Benchmarks *76*
   4.5.2 Carbon leakage list *77*
   4.5.3 Update of the carbon leakage list and benchmark values for Phase 4 *78*
   4.5.4 The correction factor *79*
   4.5.5 The New Entrants Reserve *80*
   4.5.6 State aid to compensate for the cost of carbon passed through in electricity prices *80*
4.6 Fairness and aspects of solidarity *81*
4.7 The use of auction revenues for low-carbon innovation and climate policies *83*

FIGURES

TABLES

FOREWORD

As the world's leading climate scientists tell us in the recent Special Report "Global Warming of 1.5 °C" by the Intergovernmental Panel on Climate Change (IPCC), we must act urgently and collectively, now.

If the reader is not already convinced of the need to act, I recommend that they have a careful look at the solid scientific evidence synthesised by the IPCC before reading on. The problem of climate change will continue to worsen for as long as the concentration of man-made greenhouse gases in the atmosphere continues to increase.

In Europe, we can be proud of what we have achieved so far. We have the world's most ambitious climate and energy objectives for 2030, as well as the necessary legislation to achieve them. This shows citizens that climate and energy issues are at the top of the European Union's agenda and that this priority is making real progress in Europe. To our international partners, we show that the European Union is leading by example and that we turn our pledges into action.

Now is also the moment to look ahead to 2050. We will have to further scale up our policies beyond 2030. The crucial thing is that we can only do it as part of a deep transformation of our model of economic development, one that delivers climate neutrality, prosperity and fairness for European citizens – a model that is fair for our citizens and balanced for our industries.

Far from asserting how things should be done, this book merely outlines how they are currently being done in Europe. It explains the multiple measures and instruments that have become part of Europe's overall policy.

It also describes how the EU's climate policy is searching for a new balance between public intervention and the role of markets. In this respect, fairness and effectiveness are major themes of this book. Some vulnerable communities already feel the brunt of changes to their climate, resilience and security, whereas others have until now experienced fewer impacts. Some people and companies can afford to be early adopters of new technologies, while others lack the resources to do things differently. We are all in this together, and policies must address questions of fairness as well as effectiveness.

Climate change is much more than just an environmental problem, and climate action is perhaps the defining challenge of our time. We must embark on a process of transformation with a much greater sense of urgency than I see today. We have a little time left to stabilise climate change and fulfil the goals of the Paris Agreement. We have not yet run out of time – but we cannot afford to hesitate any longer.

Words and declarations of intent need reinforcing through actions by all countries and at all levels of governance. A transformation on this scale requires an open and inclusive debate. We all need to engage widely with citizens and civil society across Europe to reach a common understanding on the way forward.

Miguel Arias Cañete
European Commissioner for Climate Action & Energy

ACKNOWLEDGEMENTS

The authors and editors would like to thank especially Ilona Billaux-Koman, Hans Bergman, François Dejean and his team at the EEA, Filip François, Tom Howes, Anna Johansson, Louisa Kelly, Jan Wouter Langenberg, Tsveti Natcheva and Alessandra Sgobbi for their contributions, suggestions and support. We are indebted to many others for help and the correction of errors and omissions, with any that remain being the sole responsibility of the authors and editors. Sincere thanks also go to Ann Mettler, Head of the European Political Strategy Centre at the European Commission, who allowed the editors time to take this book forward.

CONTRIBUTORS

Jos Delbeke

Jos Delbeke is a Professor at the European University Institute in Florence and at the KU Leuven in Belgium. He was Director-General of the European Commission's Directorate-General for Climate Action from its creation in 2010 until 2018. He holds a PhD in Economics (Leuven). Delbeke was closely involved in setting the EU's climate and energy targets for 2020 and 2030, and in developing EU legislation on the Emissions Trading System, cars and fuels, air quality, emissions from big industrial installations and chemicals (REACH). As an economist, he underlined the role of market-based instruments and of cost-benefit analysis in the field of the environment. He has been responsible for developing Europe's international climate change strategy and was for several years the European Commission's chief negotiator at the UNFCCC Conferences of the Parties. In that capacity, he played a key role in the EU's implementation of the Kyoto Protocol and in the negotiations of the Paris Agreement. In 2015, he was named Public Sector Manager of the Year (Flanders). In 2018, he received the title of Grand Officer of the Belgian Crown and the Gold Medal of the Belgian Royal Academy of Science.

Christian Holzleitner

Christian Holzleitner is currently the Head of Unit responsible for finance for innovation and land use at the European Commission's Directorate-General for Climate Action. Previously, he worked as Assistant to the Director-General for Climate Action, covering all issues related to EU and international climate policy, as well as previously at the Directorate-General for Competition in the area of state aid for services of general economic interest in the postal, transport and health sectors. Before joining the European Commission he worked as a senior manager with KPMG Germany on international transfer pricing. He is an economist and holds a PhD from the University of Linz (Austria).

Damien Meadows

Damien Meadows is the Adviser on European and International Carbon Markets at the European Commission's Directorate-General for Climate Action. Before this, he was Head of the Directorate-General's Unit for the International Carbon Market, Aviation and Maritime. He is a solicitor of the High Court of England and Wales. Before joining the European Commission in 2001, he worked for the UK government and in private practice, as well as with the United Nations Climate Change Secretariat.

Philip Owen

Philip Owen has been Head of Unit for climate finance, mainstreaming and the Montreal Protocol in the Directorate-General for Climate Action since January 2016. He has been with the Directorate-General for Climate Action since 2010, prior to which he worked as Head of Unit in the Directorate-General for Environment (2002 to 2010). Earlier in his career he worked in the Directorates-General for Competition and then Regional Policy. He joined the European Commission in 1990, having spent 12 years with KPMG in the private sector.

Alex Paquot

Alex Paquot is the Head of Unit for road transport in the European Commission's Directorate-General for Climate Action. He was previously the Head of Unit for monitoring, reporting and verification and served as Assistant to the Director-General in the same Directorate-General. Previously he worked in the Directorate-General for Environment on industrial emissions and waste management policies. Before joining the Commission in 2004, he worked for the French Ministry of Environment. He holds an engineering degree from the École des Mines in Paris.

Artur Runge-Metzger

Artur Runge-Metzger is Director responsible for climate strategy, governance and emissions from non-trading sectors in the European Commission's

Directorate-General for Climate Action. He has been responsible for the preparation of the 2050 long-term strategy, central parts of the legislation implementing the 2030 climate and energy framework for example effort sharing, land use, governance, new CO_2 emissions standards for cars, vans and trucks, as well as the implementation of domestic funding instruments including the Innovation and Modernisation Funds. From 2003 to 2015, he was one of the EU's lead negotiators in the UN climate negotiations. He served as Vice-President of the UNFCCC Bureau in 2010–2012, and he was, in 2013–2014, co-chair of the ad hoc working group preparing the Paris Agreement. Before this he spent nine years in international cooperation as Head of Operations in the European Commission's Delegation to Bosnia and Herzegovina in Sarajevo and as rural development adviser in the Delegation to Zimbabwe, as well as in the Directorate-General for Development in Brussels. He started his professional career in 1985 at the University of Göttingen, where his main scientific and lecturing topics were natural resources economics and development economics. His academic research included extensive fieldwork in rural West Africa. He holds a doctoral degree in agricultural economics from the University of Göttingen.

Yvon Slingenberg

Yvon Slingenberg is Director responsible for international climate negotiations in the European Commission's Directorate-General for Climate Action. Prior to this, she was Senior Adviser in the Cabinet of European Commissioner for Climate and Energy Miguel Arias Cañete. She has been Head of the European Commission's Units for Implementation of the EU ETS and of the Chemicals Policy Unit. She joined the Commission in 1993 and has a degree in international law (specialisation in environmental law) from the University of Amsterdam.

Tom Van Ierland

Tom Van Ierland is Head of Unit for strategy and economic assessment in the European Commission's Directorate-General for Climate Action. He is closely involved in the development of the overall climate change policy framework and the economic modelling underpinning it. He has broad experience both in the international negotiations on climate change as well as in the development of EU climate policies. He has worked for the European Commission since 2006, prior to which he started his career in Belgian public administration. He holds degrees in Applied Economics, Environmental Economics and Computer Sciences from the University of Leuven and University College London.

Stefaan Vergote

Stefaan Vergote is currently an Adviser on mitigation strategies, research and innovation in the European Commission's Directorate-General for Climate Action. He was previously Head of Unit for Economic Analysis and Financial Instruments at the European Commission's Directorate-General for Energy, as well as Head of Unit for Strategy and Economic Analysis in the Directorate-General for Climate Action. He holds a degree in electromechanical engineering and a postgraduate degree in environmental management from the University of Leuven.

Peter Vis

Peter Vis retired from the European Commission in 2019. He was previously an Adviser in the Commission's in-house think-tank, the European Political Strategy Centre. Over his career he worked in several Directorates-General of the European Commission. He was the EU Visiting Fellow at St Antony's College, University of Oxford, for the academic year 2014–2015. From 2010–2014, he was Head of Cabinet to Connie Hedegaard, European Commissioner for Climate Action. Before joining the European Commission in 1990, he worked for the UK's tax and customs authority (HM Revenue and Customs). He has a degree in History from the University of Cambridge.

Peter Wehrheim

Peter Wehrheim is a Member of the Cabinet of Phil Hogan, European Commissioner for Agriculture and Rural Development. Between 2010 and 2018, he was Head of the Unit responsible for forestry and agriculture in the context of climate mitigation policies and for climate finance in the Directorate-General for Climate Action. Prior to joining the Commission, Peter held a temporary Professorship for Agricultural and Economic Policy at the University of Bonn and was a Senior Research Fellow at the Department of Economics, University of Maryland. He also worked as a policy consultant for the FAO, the World Bank and other international development institutions. He obtained a PhD in Agricultural Policy from the University of Giessen.

Jake Werksman

Jake Werksman is Principal Adviser to the European Commission's Directorate-General for Climate Action and a lead negotiator for the European Union in the Paris Agreement process. He has held posts at the World Resources Institute, the Rockefeller Foundation, the United Nations Development Programme and the Foundation for International Environmental Law and Development (FIELD) in London. He has lectured in international environmental and economic law at the graduate level, most recently at the

New York University Law School and the Georgetown University Law Center. He holds degrees from Columbia University (AB), the University of Michigan (JD) and the University of London (LLM).

Peter Zapfel

Peter Zapfel is Head of Unit for ETS policy development and auctioning in the European Commission's Directorate-General for Climate Action. In one of his previous assignments, he was Assistant to the Director-General, responsible for policy coordination and economic assessment. He holds academic degrees from the University of Business and Economics in Vienna, Austria, and the John F. Kennedy School of Government at Harvard University.

1

HAVE 25 YEARS OF EU CLIMATE POLICY DELIVERED?

Jos Delbeke

Introduction

In line with its multilateral tradition, the European Union (EU) developed its climate policies with a view to meeting its commitments in the context of the United Nations (UN). The ratification of the Kyoto Protocol by the EU triggered the question of how the EU was to deliver the 8% emissions reduction it had committed to. It took a decade before the EU succeeded in putting a price on carbon. The agreements by the Heads of State and Government on a common target for the EU to reduce emissions for 2020 and later for 2030 were real breakthroughs. It made the elaboration of a comprehensive climate policy at the EU level possible.

1.1 The world is on a most worrying path

The facts about climate change are not at all promising. An indisputable change has been taking place in the climate system since the industrial revolution, one which has become much more pronounced since the 1970s. The concentrations of carbon dioxide (CO_2) in the atmosphere keep increasing, and this has led to a global warming of approximately 1°C increase compared to pre-industrial levels.

Solid scientific evidence has been offered by the Intergovernmental Panel on Climate Change (IPCC), based on contributions from the most qualified scientists from all over the world. In addition to the thousands of pages of

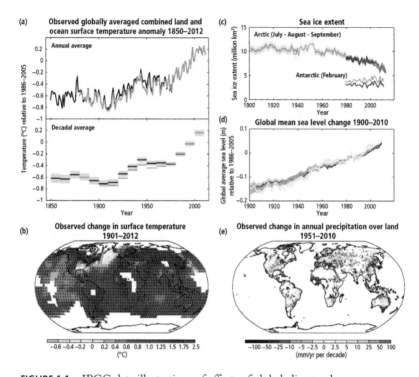

FIGURE 1.1 IPCC data illustrations of effects of global climate change

Source: IPCC 5th Assessment Report, Synthesis Report, November 2014[1] (see footnote for full citation)

scientific literature reviewed by the IPCC's studies, the IPCC also regularly produces "Synthesis Reports" with summaries for policy makers agreed by consensus (see Figure 1.1).

The fifth and latest Assessment Report of the IPCC, published in 2014, summarises the scientific insights to date. We now know with a high degree of certainty that the planet is warming at an unprecedented speed, measured over recent decades and compared to past millennia. Humans are the cause of this global warming, mainly due to rising CO_2 concentrations related to the very high consumption of fossil fuels. The impacts of climate change will be felt all over the globe in different ways and will lead to adverse impacts for humans and the economy, as well as on natural systems. Scientists also tell us that it is possible to contain the worst impacts of climate change, provided we keep global temperature increase below 2°C, and if possible, to 1.5°C,[2] compared to pre-industrial times.

The EU economy, individual citizens and society at large are already feeling the significant impact of climate change. Floods, rainfall patterns and forest fires are already happening more frequently than before, resulting in the loss of lives and damage to property and infrastructure. The extent of sea level rise may be limited to date, but the forecasts are not good: a complete melting of the huge Greenland ice sheet – even if occurring over a very long period – would make sea levels rise by around seven metres. The consequences of an expanded desertification in the Mediterranean and Africa will burden economic development, which could further boost migration pressures towards Europe. Even if the consequences of climate change will be more acute in vulnerable developing countries, the impact is likely to be considerable for Europe as well.

The IPCC also calculated a "carbon budget" related to the 2°C temperature increase limit and indicated how much is left for the future. The result represents a real policy challenge: roughly two thirds of the carbon budget compatible with the 2°C limit has already been used. However, the world population is still growing, and a continued increase is forecast for some decades.[3] The developing world is aspiring to reach income levels comparable to those of developed countries. As a consequence, in order to limit climate impacts, those with the highest *per capita* greenhouse gas emissions – and these are mainly the developed countries – must reduce their emissions earlier and very significantly. New technologies and new behavioural patterns will have to be established everywhere so as to bring down the average emissions per person on the globe to less than 2 tonnes of CO_2 per year and eventually to balance the remaining emissions with removals, achieving greenhouse gas neutrality.

All this implies that the goal of respecting the goals of the Paris Agreement to remain "well below 2°C by 2050 and pursuing efforts to 1.5°C" becomes more challenging by the day. The international community agreed to act together in 1992 at the World Summit in Rio de Janeiro through adopting the UN Framework Convention on Climate Change. Since then, many actions have been undertaken, such as the Kyoto Protocol, adopted in 1997, and more recently the Paris Agreement of 2015. The world has now agreed to step up action in all countries, also in the so-called emerging economies. The Paris Agreement has entered swiftly into force and all efforts should now be concentrated on the implementation of the commitments made. These commitments differ across the globe, and success in honouring them will also vary, but most importantly we must intensify action everywhere as a matter of urgency.

The most worrying development that has occurred since 2015 is that US President Donald Trump has announced the intended withdrawal of the

United States from the Paris Agreement. While such a withdrawal cannot be fully implemented before November 2020 (four years after the Paris Agreement came into effect), this signal of disengagement from global efforts by the world's richest country is a major blow. Fortunately, significant climate policies continue to be deployed by individual US states and cities, with many businesses also committed to contributing to climate action. Similarly, energy market developments, such as the falling costs of renewable energy, improvements in energy efficiency and coal-fired power generation not being competitive with shale gas, point towards continuous progress being made in reducing the greenhouse gas emissions of the power sector in the US.

Conclusion: Scientific evidence on climate change is unequivocal: man-made emissions of greenhouse gases have been accumulating in the atmosphere since the Industrial Revolution. Only a small window of opportunity is left to avoid significant damage to humans, nature and the economy. The international community has decided to act together and to develop policies to limit climate change to "well below 2°C, and pursuing efforts to 1.5°C".

1.2 The EU reduced its greenhouse gas emissions by 22% since 1990

Following adoption of the UN Framework Convention on Climate Change, the EU decided in 1993 to gather the necessary data on its emissions of greenhouse gases. It was the start for the annual publication of a factual and impartial report, produced by the European Environment Agency (EEA), indicating how Europe's greenhouse gas emissions evolve and how much sectors emit in each of the Member States.

The good news is that the EU has been reducing its emissions consistently since 1990, the base year of the UN Framework Convention on Climate Change, the Kyoto Protocol and the Paris Agreement (Figure 1.2). In 2017, the EU had already achieved a 22% reduction, which is higher than the target it set itself of a 20% reduction below 1990 levels by 2020.

Heads of State and Government decided in 2014, in readiness for what was to become the Paris Agreement, that the EU would increase the ambition of its greenhouse gas emissions reductions to at least 40% below 1990 levels by 2030. While this will require more action, the bigger challenge will be to prepare the world for net zero emissions in the second half of the

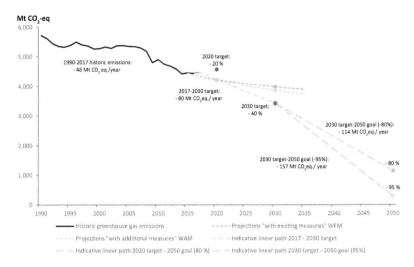

FIGURE 1.2 EU greenhouse gas emissions 1990–2050: historical record and challenges ahead

Source: EEA trends and projections report 2018, EEA GHG dataviewer 2018, EEA GHG projections dataset 2018

century, as stated in the Paris Agreement. The recent IPCC Special Report on 1.5°C reviews the available scientific literature. It concludes that far-reaching and rapid transitions in all sectors around the world will have to be made. Subsequently, the European Commission has indicated that it is possible with currently known technologies to realise such a deep transition towards a climate-neutral Europe by 2050.[4]

In the EU, carbon dioxide (CO_2) represents more than 80% of total greenhouse gas emissions. This release of CO_2 is connected with the extensive use of fossil fuels in power generation, in industry and transport, and these sectors have therefore been the main targets for policy intervention. The other greenhouse gases are methane (CH_4), nitrous oxide (N_2O) and fluorinated gases (F-gases). They originate mainly from the agricultural, chemical and waste sectors. These non-CO_2 gases are produced in smaller volumes but are more potent global warming gases, so they have a correspondingly greater impact on the climate system.

The EU's emissions trend was significantly curbed between 2005 and 2017 (see Table 1.1). This represents a step-change and will make more ambitious reductions in the future both feasible and affordable. Between 1990 and 2005,

TABLE 1.1 EU greenhouse gas emissions by sector, 1990–2030 (in CO_2 equivalent and as percentage of 1990 emissions)

EU greenhouse gas emissions by sector, 1990–2030
(2030 projections "with existing measures")

(in Million tonnes of CO_2 equivalent)					*Percentage variations since 1990*		
	1990	2005	2017	2030	2005	2017	2030
Energy supply	1869	1713	1276	1053	92	68	56
Energy use in Manufacturing	841	636	483	459	76	57	55
Industrial processes and product use	517	466	379	340	90	73	66
Transport	787	976	946	887	124	120	113
Other energy use	854	794	663	555	93	78	65
Agriculture	542	434	432	432	80	80	80
Waste	236	200	136	99	85	58	42
International aviation	69	131	150	164	190	217	238
Total	5715	5350	4465	3989	94	78	70

Source: Author based on EEA data

annual reductions were limited to less than half a percentage point per year. An important change came after 2005, when the EU began its integrated climate and energy policy; compared to 1990 the emissions index declined from 94 to 78 in fewer than 12 years. The annual emission reduction between 2005 and 2017 jumped from less than 0.5% to 1.5% per year. Continuing towards the targets set for 2030 will require a further annual increase of approximately 2%. In early 2019, Member States presented their draft National Energy and Climate Plans for this higher ambition level, as past policies for 2030 only lead towards an emissions index 70 instead of 60 (100 = 1990), which would correspond with the "at least 40% target" confirmed by the EU's ratification of the Paris Agreement. Covering this emissions reduction gap will require a significant strengthening of current policies.

This curbing of the EU's greenhouse gas emissions trend is primarily the result of evolutions in the power and manufacturing sector. The combination of policies related to energy efficiency, renewables and carbon pricing seemed effective. On the other hand, the transport sector and in particular, aviation increased its emissions, thereby neutralising some of the progress made elsewhere. Emissions from waste have been considerably reduced even if this sector covers only a minor proportion of the EU's emissions. The next phase of political action until 2030 therefore has to focus more on the curbing of greenhouse gas emissions from transport. The required technologies are increasingly available. The EU has adopted ambitious new CO_2 standards for new passenger cars in order to improve their performance by 37.5% over a period of nine years from 2021–2030.

The amount of carbon stored in the EU's forest, soil and vegetation is not included in the previous data. This so-called carbon "sink" has been more or less stable and was estimated to be around 230 million tonnes of CO_2-equivalent in 1990 and 304 million tonnes of CO_2-equivalent in 2015. This is primarily the result of sustainable forest management and afforestation. In the future, much more policy attention also needs to be given to the fixing of carbon in the soil by means of proper incentives provided by the EU's Common Agricultural Policy, for example. Equally, an increased combustion of biomass should not lead to additional emissions. For these reasons, a systematic accounting tool has been developed[5] (as further explained in Chapter 8).

Countries and continents have very different emission profiles when it comes to greenhouse gases. In Europe, the main contributor is CO_2 emissions from fossil fuel use coming primarily from power generation, transport and manufacturing. In other countries outside Europe, the major concern may be tropical deforestation or methane emissions from cattle. This variation pleads in favour of a policy approach allowing for considerable flexibility. Not only do the emissions differ significantly; income and wealth are also unevenly distributed across the globe. The "bottom-up" approach as enshrined in the Paris Agreement through the so-called "Nationally Determined Contributions" captures the need for such policy flexibility.

> Conclusion: The EU has reduced its emissions in 2017 by 22% compared to 1990. The major contribution to emissions reduction came from the power and manufacturing sectors, while transport emissions kept growing.

1.3 The EU decoupled its emissions from economic growth

No "silver bullet" or single technology is able to bring down greenhouse gas emissions across the economy. Consequently, multiple measures are required, and they should be designed in a way that creates as many synergies as possible. After the EU's ratification of the Kyoto Protocol in 2002, the

EU introduced a number of policies. It began with the adoption of the EU Emissions Trading System (EU ETS) Directive in 2003, which established a carbon market in Europe from 2005, covering emissions from the power and manufacturing sectors. For the sectors of the economy not covered by the EU ETS, such as transport, buildings and agriculture, the EU agreed legally binding targets per Member State.

A most notable result of this intensive policy activity is that the EU succeeded in bringing down its greenhouse gas emissions without sacrificing economic growth. Over the period from 1990–2017, the EU increased its GDP by 53% while at the same time reducing its emissions by 22% (Figure 1.3). As a result of this successful decoupling, the greenhouse gas emissions' intensity per unit of GDP was halved between 1990 and 2017. The continuing reduction of greenhouse gas intensity over more than two decades demonstrates also that progress in terms of decoupling has been made notwithstanding the irregularity of economic cycles.

It is sometimes claimed that this decoupling is unduly flattered by the accounting methodology used. This methodology is based on the "direct" emissions approach, as adopted by the UNFCCC and confirmed in the Paris

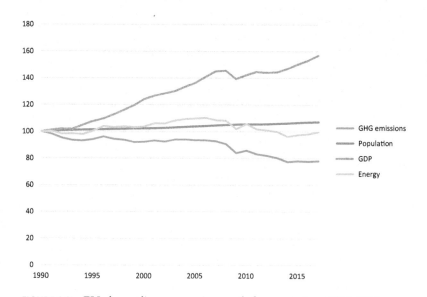

FIGURE 1.3   EU: decoupling economic growth from emissions 1990–2017

Source: EEA, based on data from EEA (GHG emissions), Eurostat (energy and population) and DG ECFIN, European Commission (GDP)

Agreement. This approach makes all countries directly responsible for the emissions emitted on their territory. The universal participation of Paris Agreement reinforces the inclusiveness of this approach. Policies that would indirectly address emissions coming from other countries could be perceived as a breach to this territoriality principle and to national sovereignty more broadly.

The alternative would be an "indirect" approach whereby emissions are accounted for in relation to the consumption of goods within a country, whether imported or manufactured. Behind this assertion is the argument that the EU is simply importing more carbon-intensive products and is thereby "exporting" its carbon emissions to the rest of the world. While this may be happening to some extent, the overall result is counterbalanced by the fact that the EU's exports are relatively efficient in terms of carbon content and positively contribute to bringing down emissions in the rest of the world. Modelling done for the European Commission suggests that the EU has contributed to the decarbonisation of third countries by raising the energy efficiency of its own economy and exports; EU exports have been estimated to reduce global emissions in 2016 by a little more that 200 million tonnes of CO_2-equivalent compared to a situation where EU exports would be produced locally in the importing countries.[6] One can very well conclude that the decoupling of emissions from economic growth is a fundamental feature of the EU's economy and should be continued in the future. It shows that climate policy is not undermining economic growth; rather, what is important is the quality and the embedded technology of that growth.

The two main drivers behind the decoupling of economic growth and emissions have been the reduced energy intensity of the economy, and the reduced carbon intensity of energy production (Figure 1.4). Since 2000, these two factors could compensate for the emissions influence of an increase in welfare (GDP *per capita*) and increased population. It underlines the major changes that were happening in the energy sector, not least in power production. Between 2008 and 2015, which includes the worst years of the economic recession following the financial crisis from 2008, CO_2 emissions from fossil fuel combustion fell by 14.4%.[7] These reductions were impacted by the policies both to curb energy demand and to decarbonise energy supply.

These developments have obviously had an impact on the position of Europe compared to the rest of the world. The EU's share of global emissions has been falling continuously, partially because other parts of the world, especially the emerging economies, have been emitting much more.

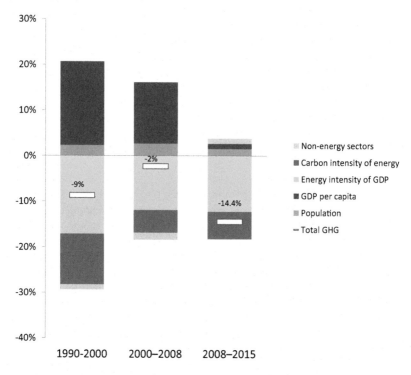

FIGURE 1.4 Analysis of key trends and drivers in greenhouse gas emissions in the EU, 1990 and 2015[8]

Source: EEA "Analysis of key trends and drivers in greenhouse gas emissions in the EU between 1990 and 2015"

The combined G20 countries were responsible for three-quarters of global greenhouse gas emissions in 2012, while the EU's share of global greenhouse emissions was 10% and falling. China's share was over 26% in 2012, followed by the US (13.5%), India (7%) and Russia (5%).[9] These comparative numbers will, of course, have evolved considerably since then.

In 2015, CO_2 emissions *per capita* were 6.9 tonnes CO_2 in the EU and have shown a steady reduction since 1990 (see Figure 1.5). At the same time, in China, CO_2 emissions *per capita* have risen significantly since 2002. These emissions in China reached the EU's level in 2012, and since then have continued to grow, reaching 7.7 tonnes CO_2 in 2015. Although declining, *per capita* emissions of the US are very high compared to the global average and are more than twice those of the EU. Levels *per capita* in India, for example, are still very low but are steadily increasing.

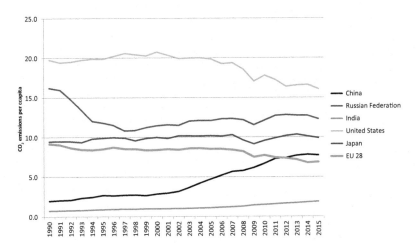

FIGURE 1.5 Global *per capita* CO_2 emissions from fossil fuel use and industrial processes, 1990–2015

Source: Emission database for global atmospheric research (EDGAR), 2016 update: http://edgar.jrc.ec.europa.eu/

> Conclusion: The EU reduced its greenhouse gas emissions while maintaining economic growth. This decoupling took place in an incremental but steady manner, thanks to the integrated climate and energy policy that encouraged low-carbon technology and investment.

1.4 The five cornerstones of EU climate policy

The EU is a transnational regional organisation composed of 28 sovereign Member States with highly diverse characteristics in terms of wealth, income, industrial performance and natural endowment. It is therefore not surprising that the public debate on climate change differs significantly amongst them, as is the political preparedness to take action. It was key to ensure that all Member States became part of the policy solution and felt their specific concerns were adequately addressed in the policy architecture.

It was imperative for the EU to embed its emerging climate policy firmly into the multilateral context of the UNFCCC. It was necessary to point at the global nature of the problem, as much as to the global dimension of the policies required. It was also an important way to demonstrate to the emerging economies, now responsible for the bulk of the world's greenhouse gas emissions, that their climate policies can deliver important benefits to them as well.

1.4.1 A vision endorsed by the highest political level

While Europe recognised the magnitude of the climate challenge, a major question was to determine its fair share of that global task, both technically and politically. Europe recognised its significant historical contribution to the greenhouse gases accumulated in the atmosphere. However, it was equally important to search for a maximum emission reduction at reasonable cost as it would be of utmost importance to preserve the collective "willingness to pay." Even if we all benefit from the long-term results of the policy, there are nevertheless short-term additional investments involved. This raises inevitably political questions, even by those for whom the need to act is absolutely accepted.

In Europe, a clear endorsement at the highest levels of a long-term policy vision turned out to be critical. As of 2005, the European Council – comprising of the EU's Heads of State and Government – had discussed the issue several times. In 2007, the leaders formally agreed on an "independent commitment to achieve at least 20% reduction of greenhouse gas emissions by 2020 compared to 1990." They even offered to increase this commitment to 30% by 2020 if other developed countries would commit themselves to a comparable emission reduction. This was clearly an invitation to the US to join in the overall efforts. The climate commitment was accompanied by specific energy objectives, including a binding target to increase renewable energy's share in final energy consumption to 20% by 2020 (from about 8.5% in 2005) and an indicative target to reduce energy consumption by 20% in 2020.[10]

On this basis, the Commission developed a package of legislative proposals that was finally agreed one year ahead of the UNFCCC conference in Copenhagen in December 2009. While the European Council's guidance was innovative institutionally, by reason of the level of detail it gave in its orientations, the European Parliament accepted this innovation, knowing that it was bound to have a decisive say and opportunity to assert its position when deciding upon the individual pieces of the legislation.

Five years later, in 2014, the first permanent President of the European Council[11] Mr Herman Van Rompuy continued the role of encouraging climate policy through sealing key decisions at the level of Heads of State and

Government. Based on the technical preparation by the Commission, not least in its "Roadmap for a competitive low-carbon economy,"[12] the European Council formally decided on the "at least 40% greenhouse gas reduction" to be achieved by 2030. This decision constituted the EU's political and diplomatic contribution in the run-up to the Paris Climate Conference at the end of 2015, and it had a critical influence on the ambitions set out by the US and China a few weeks later.

These decisions on the long-term vision of its climate policy created a precedent on how to deal with important new challenges in the EU. According to the Lisbon Treaty, the Council of Ministers and the European Parliament adopt legislation together based on a proposal prepared by the European Commission. An innovative institutional practice was created whereby the European Council, which brings together the Heads of State and Government of the 28 Member States, adopted a general policy orientation before specific legislation was prepared.

This original way of getting the new EU institutions to work out strategic policy questions in the evolving context of the Lisbon Treaty hinged on the cooperation between key persons. The Heads of Cabinet of the Presidents of the Council and the Commission, and their respective Secretary-Generals, were working hand in glove to get this new policy area on the rails. It allowed for bridging the gap between technical and political knowledge, which is so critical to deliver solid and lasting outcomes.

The heart of the strategic vision adopted by the Heads of State and Government was the setting of clear and simple targets. Over time, these were gradually tightened, from 8% to 20% and later to 40% emissions reductions. Now that the objective for 2030 has been translated into legislation, the next critical debate is likely to shift to the ones for 2040 and 2050. Further negotiations at the international level on more climate ambition would benefit greatly from sharing this experience of creating a common vision amongst 28 sovereign States, while finding solutions to overcome their social, economic and political differences.

Conclusion: Heads of State and Government, acting together in the European Council on the basis of consensus, gave clear guidance on EU-wide targets for greenhouse gas reductions, renewable energy and energy efficiency with respect to 2020 and 2030. This practice significantly facilitated the subsequent adoption of specific EU legislation.

1.4.2 Putting an explicit price on carbon

In parallel with the decision to ratify the Kyoto Protocol, the EU policy-implementation phase started with the creation of the European Climate Change Programme. This was an extensive consultation exercise with stakeholders from civil society, the business sector and national authorities. As greenhouse gases are emitted by so many different economic activities, the consultation process ended up with a long list of possible actions, after which a process of priority setting was needed.

The first issue was to focus on policies for which the European Commission was best placed to propose legislation at the European level. In some areas, national action is difficult to put in place because of potential distortions of competition. Since the 1980s, it has become increasingly clear that adopting rules at European level was an effective and practical way to overcome such distortions.

The conclusion of the European Climate Change Programme was that several layers of government must endeavour to work together, and that the European level was best placed for action where the single market perspective created a strong value added. It was decided to act in a harmonised manner in the fields of energy-intensive and manufacturing industry, electricity and heat production. This enabled the harnessing of economies of scale provided by a market of more than 500 million consumers.

In the 1990s, economists launched a debate on "pricing economic externalities" and how to put this into practice. Prices are a very effective way of transmitting information through the economy and influencing behaviour, right down to the levels of individual producers and consumers. This can be achieved through taxes, or alternatively, through the setting of overall limits to pollution levels ("a cap"). In other words, economic instruments can work either directly through setting prices for polluting (e.g., through taxes), or through defining quantities of pollution allowed (e.g., through capping pollution levels). Businesses expressed a clear preference for carbon pricing instead of detailed technical regulations.

The initial proposals at EU level were for pricing through taxes. In 1992, the Commission made a proposal for a combined carbon and energy tax. Under the EU Treaty, this required unanimity. After almost a decade of difficult negotiations, the tax approach was abandoned, in particular due to the reservations some Member States had in allowing the European Union more say in taxation policy (notwithstanding that some indirect taxes were already regulated at European level, most notably Value Added Tax and Excise duties). Moreover, taxes seemed difficult to handle politically almost everywhere, as people easily feel over-taxed. The Minister of Finance may also look more for fresh revenues, while the climate and/or environment minister looks first at curbing pollution. These two interests are not necessarily easy to combine. The European debate on economic instruments shifted from taxation to cap setting and emissions trading. The US was the first to demonstrate in practice the advantages that an instrument such as "cap-and-trade" could offer. Led by their policy experience on reducing sulphur and NO_x emissions, the US pushed successfully for its introduction into an article of the Kyoto Protocol, against the wishes of the Europeans and others. Setting a limit on the total amount of emissions could ensure emissions reductions, and the trading of "permits," or "allowances," could offer cost-efficiency. Possible legislation for emissions trading at EU level would fall under the Environmental Chapter of the Treaty of Rome (as then amended by the Maastricht Treaty), and therefore decisions could be made through qualified majority voting in Council. In such a manner, the political and institutional stalemate that had blocked progress on EU-level carbon and energy taxation could be overcome.

In 2003, it was decided to start the EU Emissions Trading System (EU ETS), covering all major actors in the field of power and manufacturing. More than 11,000 industrial installations and subsequently airlines across Europe were covered by one EU-wide cap on emissions, rather than having distinct national targets for EU ETS-covered sectors.[13] The EU-wide approach has effectively ensured that abatement is achieved where the costs are the lowest, without creating distortions of competition between large industrial installations across Europe. The EU ETS gradually established a truly internal market for carbon emissions, applicable in a harmonised manner from northernmost Sweden or Finland to the south of Italy or eastern Romania. It developed in synergy with the gradual establishment of the internal market for electricity.

A debate continues amongst economists whether the incentives created by the EU ETS have been solid enough to generate significant emission reductions. However, between 2005 and 2016 emissions under the EU ETS decreased by 26%, i.e., more than the average reductions made across the EU and never exceeding the cap. More than the price of carbon, the real emission reductions matter most.

Conclusion: The EU ETS created an explicit price on carbon emissions and reduced emissions by 26% between 2005 and 2017. The system is cost-effective as it offers considerable flexibility to companies. It creates incentives to save energy and to innovate in low-carbon technologies, such as renewable energy.

1.4.3 The need for a comprehensive policy approach

Putting a price on carbon is of capital importance but is not sufficient in itself. It is most useful in sectors driven by a strong economic rationale, such as in private business. Moreover, the market does not take into account distributional issues or other political realities that matter.

Economists have been struggling with the issue of double regulation, as alongside carbon prices other regulations, such as on renewable energy and energy efficiency, may exist in parallel. The real world is, unlike theoretical models, always subject to multiple regulations, and some of them may partially overlap. The important question is whether these regulations work in the same direction or undermine one another. The EU ETS has definitely been a solid instrument to realise low-cost emission reductions, working as a complement to legislation on renewable energy that worked with much higher implicit carbon prices. To the extent that separate renewable energy legislation or energy efficiency standards could create a surplus in the carbon market, the newly created Market Stability Reserve neutralises that effectively in the future (see Chapter 4).

While the EU ETS may cover some 45% of emissions, many small emitters such as households, transport users and agriculture generate the remaining 55%. Dealing satisfactorily with these actors can hardly be managed through a harmonised European approach; hence, these emissions are shared out between the Member States in the form of legally binding targets. Based on economic modelling, a cost-effective reduction pathway has been calculated for the overall EU target of a 40% reduction in greenhouse gases by 2030 compared to 1990 between the sectors covered and those not covered by the EU ETS. It was therefore decided to reduce the emissions under the EU ETS by 43% in 2030 compared to 2005, combined with a collective reduction target for the sectors outside the EU ETS across all Member States of 30% in 2030 compared to 2005.

For the first time however, the EU target also accounts for emissions coming from the agriculture and forestry sectors, often called the "Land Use, Land Use Change and Forestry," or "LULUCF" sectors. The EU's 2030 target of at least a 40% reduction has the particularity, however, that no net increase in emissions are to come from these agriculture and forestry sectors, which is technically referred to as the "no-debit rule."

EU climate policy is comprehensive as the EU ETS, the Effort Sharing Regulation defining targets for the Member States for the non-EU ETS sectors and the LULUCF sector together cover the totality of Europe's greenhouse gas emissions. There are only very limited gateways allowed between

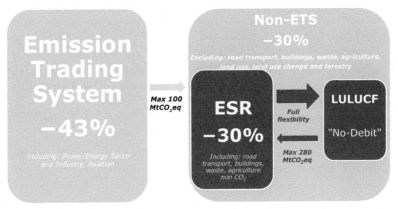

EU commitment: at least −40% domestic GHG emissions reductions by 2030

FIGURE 1.6   EU climate policy design

Source: European Commission

those three pillars: a maximum of 100 million tonnes between the EU ETS and Effort Sharing parts and a maximum 280 million tonnes between the Effort Sharing and LULUCF parts. In order to make sure the targets are delivered, the EU has established a harmonised emissions accounting system.

> Conclusion: The EU has developed a comprehensive policy approach whereby the EU ETS is complemented with targets for the Member States covering the emissions of households, transport users and agriculture. A third pillar of LULUCF has been added under the 2030 overall target.

1.4.4 Solidarity and fairness: addressing the distributive questions

The EU's overall policy approach has been guided by the principle of cost-effectiveness. However, this approach needed a correction for distributive impacts. Economic analysis had indeed shown that there would be considerable differences in overall investment needs if targets were to be distributed purely on the basis of cost-effectiveness (i.e., equalisation of marginal cost of abatement per Member State). Indeed, that would imply that lower-income Member States, notably in Central and Eastern Europe, would have to face higher additional investments in relative terms because of their higher energy- and carbon-intensity and their lower GDP. Asking poorer Member States to do more than richer ones just because it is more cost-effective was obviously unacceptable to the poorer States. As the *per capita* income disparity between the EU's poorest and richest Member States is more than 1:10, the EU has been ensuring that fairness is a central concept in its climate policy.

One way of guaranteeing fairness is through the differentiation of emissions targets for each Member State in sectors outside the Emissions Trading System. The effort was shared out between Member States based on national *per capita* income.[14] The result is a set of differentiated targets ranging from a 40% reduction compared to 2005 for the highest income countries to 0% compared to 2005 for countries with the lowest average *per capita* income. Similarly, on renewable energy for 2020, national targets were formulated in such a way as to promote a fair distribution between Member States. Overall, Member States with similar economic performances, often neighbouring each other, received similar targets.

Considering the uncertainties related to future economic development and to enhance cost-effective achievement of targets, further flexibility allowed Member States to transfer emission rights between themselves. In this way, countries with higher national costs could achieve their target more cheaply, and countries that overachieve their target could benefit financially. There is clearly, therefore, a link between flexibilities and fairness, as the flexibilities allow for the transfer of obligations in exchange for revenue, and the obligations are set in such a way as to give benefit to the relatively poorer Member States.

As part of the EU ETS design, a significant share of allowances has been set aside for specific purposes. Ten percent of allowances for auctioning are redistributed in favour of lower income Member States. Moreover, a proportion of allowances are set aside to create a Modernisation Fund that provides important financial support for modernisation of the energy sector in these Member States. At a price of €20 per tonne, this represents an envelope of some €6.5 billion to be used for low-carbon and energy efficiency purposes. If we take all redistributive provisions of the EU ETS in favour of the lower income Member States together, some 1.5 billion allowances are available during the period from 2020–2030. At a price of €20 per tonne, this adds up to some €30 billion or an average of some €3 billion per year.

Finally, to facilitate the transformation of specific regions towards the low-carbon world of the future, the Commission has put greater focus on climate change in the EU's regional policy instruments. After all, the EU's experience in creating alternative employment in former coal mining regions in the Benelux countries, France and Spain has been quite successful. This policy is now being reinforced in the Commission's Coal and Carbon-Intensive Regions in Transition Initiative.[15]

Conclusion: EU climate policy addresses redistributive concerns; targets are differentiated according to GDP *per capita* and lower-income Member States stand to receive some €30 billion of additional revenues through the EU ETS over the period from 2020–2030.

1.4.5 Addressing the competitiveness of EU manufacturing industry

Another major issue that needed to be addressed is the impact of the regulations on the competitive position of EU companies. From day one of the development of EU climate policy, manufacturing industries have been worried about the potential impact of policies on their competitive position. Unlike the power sector that basically produces for the European market, these companies are often dependent on world markets and global price setting. Even if these industries accepted the need for emissions reduction as a medium- and long-term objective, the potential impact on their global competitive position needed to be addressed. For that purpose, a system has been developed that seeks the right balance between the stick (the price) and the carrot (the free allocation of allowances, or support for innovation).

The "pilot phase" of the EU ETS, namely the years 2005–2008, had made one major concession, namely that the decision on the allocation of allowances was left to the Member States, with limited oversight. An added difficulty at the time was that there was little data on the past CO_2 emissions of individual operators. Unsurprisingly, Member States did not want to take any risk of undermining the competitive position of their industry, and the consequence was a significant over-allocation of emission allowances and carbon prices falling close to zero. Correction was only possible by application of a more restrictive approach to allocation and much stronger Commission scrutiny of subsequent national allocation plans. This experience showed how difficult it is to make optimal allocation decisions when getting a system off the ground, such as in the absence of good data.

Today, free allocation is subject to harmonised rules that apply to all industrial operators across the EU. It is based on technological benchmarks for different sectors of industry, of which there are 54,[16] and on considerations of how exposed industry sectors are to global competitive pressures. It was decided that up to 43% of the allowances under the EU ETS will be handed out for free between 2020 and 2030. On top of that an EU Innovation Fund, endowed with the revenues from the sale of at least 450 million allowances, has been created to support innovation projects for example to finance innovation in renewable energy, Carbon Capture and Storage, energy storage and lower-emitting industrial projects. At a price of €20 per tonne, this represents €9 billion over the period from 2020–2030, for incentivising innovation in industrial and energy companies.

> Conclusion: The competitive position of the EU's industry is addressed in the EU ETS, both through free allocation and through an Innovation Fund of some €10 billion.

Conclusion

Today, the EU's greenhouse gas emissions are 22% lower than in 1990. This was achieved without sacrifices in terms of economic growth or jobs. On the contrary, important new economic activities were created and intensive low-carbon innovation and employment can be expected in the future.

Decisions of the Heads of State and Government combined with the adoption of a comprehensive vision and specific targets for 2020 and 2030 have been of capital importance. They paved the way for realising the ambitious goals of the Paris Agreement and creating the space to move gradually to net-zero emissions in the EU by 2050. Because the EU is comprised of 28 sovereign Member States with very different economic, social, political and geographic conditions, these decisions have also turned the European continent into a laboratory for the fair implementation of climate policies, useful to the world at large.

One of the lessons learnt from the EU's experience is the importance of a gradual tightening of its overall and specific policy objectives. The target started with reductions based on 1990 numbers of 8% by 2012, 20% by 2020 and 40% by 2030. Similarly, the objectives under the EU ETS, the Effort Sharing Regulation, the renewable energy and energy efficiency Directives were gradually strengthened in the light of experience. Starting slowly gave the chance to both public and private actors to learn how to tackle things in practice. It was necessary to learn from experience, for example, how to handle free allocation better under the EU ETS, or how to rely more on wind and solar energy instead of food-based biofuels. This learning-by-doing was invaluable and reinforced the confidence that more results lie within reach. This experience may help other countries go faster in the development of their own policies.

Another useful lesson is that political debates on climate policy, given their complexity, need meticulous technical preparation. The well-established tradition within the European Commission of undertaking Impact Assessments has proven most helpful in this respect. The task of building a portfolio of climate policies may appear challenging, but it is worth sharing this practical experience to facilitate the implementation of the Nationally Determined Contributions by all countries that have joined the Paris Agreement. Having a well-informed debate informed by good analysis helps policies forward and fosters wider support.

Addressing potential problems in an open manner serves policy development in a longer-term perspective. In particular, addressing distributional consequences are important as it enables the design of accompanying policies so that those who might otherwise see themselves as potential losers now share in the benefits of success. Similarly, competitiveness pressures on manufacturing industries that are active on world markets cannot be denied as long as similar policies are not put in place by trading partners. Dealing with these issues is not easy. Nevertheless, the EU experience has proven that there is no alternative if the objective is to develop climate policies in the light of ever more ambitious emissions reduction targets. This also places climate policies at the centre of a much wider, future-orientated political agenda.

Notes

1 IPCC (2014). *Climate Change 2014: Synthesis Report. Contribution of Working Groups I, II and III to the Fifth Assessment Report of the Intergovernmental Panel on Climate Change* [Core Writing Team, R.K. Pachauri and L.A. Meyer (eds.)]. Geneva: IPCC, p. 151.
2 There were clear indications of the benefits to limit warming to 1.5°C in the IPCC's Special Report "Global Warming of 1.5 °C" of October 2018.
3 The current world population of 7.6 billion is expected by the UN to reach 8.6 billion in 2030, 9.8 billion in 2050 and 11.2 billion in 2100. www.un.org/development/desa/publications/world-population-prospects-the-2017-revision.html.
4 European Commission Communication (2018). "A Clean Planet for All – A European Strategic Long-Term Vision for a Prosperous, Modern, Competitive and Climate-Neutral Economy". COM(2018)773 final of 28.11.2018.

5 Regulation (EU) 2018/841 of the European Parliament and of the Council of 30 May 2018 on the inclusion of greenhouse gas emissions and removals from land use, land use change and forestry in the 2030 climate and energy framework and amending Regulation (EU) No 525/2013 and Decision No 529/2013/EU.
6 See pp. 263–267 of the European Commission's "In-depth Analysis in Support of the Commission Communication COM(2018)773 – "A Clean Planet for All: A European Long-Term Strategic Vision for a Prosperous, Modern, Competitive and Climate Neutral Economy". 28.11.2018. https://ec.europa.eu/energy/en/topics/energy-strategy-and-energy-union/2050-long-term-strategy.
7 Report from the Commission to the European Parliament and the Council: Progress towards achieving the Kyoto and EU 2020 objectives, Brussels (COM(2014)689 final dated 28.10.2014). http://ec.europa.eu/transparency/reg doc/rep/1/2014/EN/1-2014-689-EN-F1-1.Pdf. The figure is based on analysis of the EEA (2012). "Why Did Greenhouse Gas Emissions Decrease in the EU in 2012?" *European Environment Agency – EEA Analysis*, Copenhagen, 3.6.2014. www.eea.europa.eu/publications/why-did-ghg-emissions-decrease.
8 The decomposition analysis is based on the logarithmic mean Divisia index (LMDI). The bar segments show the changes associated with each factor alone, holding the other respective factors constant. The explanatory factors should not be seen as independent of each other.
9 The version v4.3.2 of the EDGAR emission inventory provides global emission estimates by region/country for all anthropogenic activities except the land-use, land-use change and forestry sector (including Forest fires and Savannah burning) for the three major Greenhouse Gases (CO_2, CH_4 and N_2O, summed in CO_2 equivalent using the GWP-100 metric of AR4). Further details can be found in the ESSD publication of Janssens-Maenhout, G., Crippa, M., Guizzardi, D., Muntean, M., Schaaf, E., Olivier, J.G.J., Peters, J.A.H.W., and Schure, K.M. (2017). *Fossil CO_2 and GHG Emissions of All World Countries*. Luxembourg: Publications Office of the European Union.
10 Compared to the EU's projected energy consumption in 2020, as established in the 2007 baseline scenario of the Impact Assessment for the 2020 climate and energy package (see Capros, P. L. Mantzos, V. Papandreou, N. Tasios (2008) "European Energy and transport trends to 2030 – update 2007". Publications Office of the European Union, Luxembourg. https://ec.europa.eu/energy/sites/ener/files/documents/trends_to_2030_update_2007.pdf).
11 Creation of a permanent Presidency of the European Council was an innovation introduced by the Lisbon Treaty's amendments to the Treaty of Rome (the latter being the founding Treaty of what is now the European Union).
12 Add footnote and link to the document.
13 An "upstream" mechanism whereby fuel imported and produced would be subject to a charge based on its carbon content was discarded as amounting to a tax, which would probably entail the need for unanimity among Member States in Council (and deprive the European Parliament from having a meaningful say in the design of the instrument).
14 This principle was also partially used in the definition of the renewable energy 2020 targets for each Member State. For 2030, EU Member States do not have individual renewable energy targets but must rather make National Climate and Energy Plans (see Chapter 6 for more details of the Energy Union Governance Regulation).

15 Coal – including both hard coal and lignite – is currently mined in 12 EU countries and is an important source of economic activity in coal mining regions. The coal sector provides jobs to about 240,000 people: about 180,000 in the mining of coal and lignite and about 60,000 in coal- and lignite-fired power plants. For more details see: https://ec.europa.eu/energy/sites/ener/files/crit_tor_fin.pdf.

16 52 product and two fallback benchmarks.

2

THE PARIS AGREEMENT

Jos Delbeke, Artur Runge-Metzger,
Yvon Slingenberg and Jake Werksman

Introduction

Since their emergence from the Second World War, the Member States of the European Union have consistently preferred a multilateral approach to global problems, as is reflected through the EU's support for the UN and for institutions like the World Trade Organisation (WTO). The establishment of the European Union is another manifestation of that same commitment to multilateralism.

Since the publications of the Club of Rome[1] in the 1970s and more particularly since the first Earth Summit in Stockholm in 1972, Europeans have been calling on the world to act together to halt and reverse environmental degradation. Climate change is undoubtedly the most typical of global environmental problems, for which there is no solution without international cooperation. For these reasons, European countries have worked hard to find coordinated solutions through the UN, including most recently through the Paris Agreement.

This chapter describes the origin, content and essential features of the Paris Agreement of 2015 and explains why it has commanded near universal support and participation from the international community. We describe how the Paris Agreement, nonetheless, remains an international pact whose ambitious goals can only be achieved through strong and active cooperation amongst its Parties.

In June 2017, US President Donald Trump announced his intent to withdraw the US from the Paris Agreement. Losing the world's largest economy and the second largest emitter of greenhouse gases, as has happened with other global treaties, undoubtedly undermines multilateralism, especially if it would turn into a downward spiral, for instance if others were to follow or if individual US states, municipalities and other stakeholders would stop their efforts to tackle climate change. This challenge adds to the fact that much more ambition is necessary to turn the commitments made under the Paris Agreement into policies and action on the ground. This is particularly the case for emerging economies, as their action on their rapidly growing emissions will significantly determine the climate change impacts of the future.

2.1 The UN Framework Convention on Climate Change and the Kyoto Protocol

The Paris Agreement is the third generation of international treaties designed to respond to the challenge of climate change. The first, adopted in 1992 just prior to the United Nations Conference on Environment and Development in Rio de Janeiro, is the UN Framework Convention on Climate Change (UNFCCC).[2] The Convention contains an important objective – stabilising greenhouse gas concentrations in the atmosphere at safe levels. It established the principal institutions necessary for the UN's climate regime to function, including the UNFCCC Secretariat and its governing body – the annual Conference of the Parties (COP).

Most significantly, the UNFCCC established the first international system for the national reporting of inventories of greenhouse gases and for communicating policies and measures that Parties have put in place to manage their emissions and adapt to the impacts of climate change. In compliance with its obligations under the UNFCCC, the European Union developed and submitted its first greenhouse gas inventory to the UNFCCC Secretariat in June 1996, as part of its first National Communication.

The Convention also sets out key principles intended to guide international cooperation on climate policy, including an expression of the precautionary principle which calls on governments to act when faced with threats of serious or irreversible damage, even if there is a lack of full scientific certainty about the nature of those threats. The Convention's principles also recognise that measures to address climate change should promote sustainable development, should be appropriate to the conditions of each Party and should not constitute a means of arbitrary or unjustifiable discrimination or a disguised restriction on international trade.

Importantly, the Convention calls on Parties to address climate change "on the basis of equity and in accordance with their common but differentiated responsibilities and respective capabilities." Accordingly, the Convention states that developed country Parties (like the EU and its Member States) should take the lead in combating climate change and the effects thereof.

In 1992, when the Convention was adopted, the principle of "common but differentiated responsibilities and respective capabilities" was made operational through the Annexes of the Convention. Annex I listed those industrialised Parties considered "developed." These countries undertook to aim to stabilise their emissions of greenhouse gases at 1990 levels by 2000. The richest of these (the then members of the Organization for Economic Cooperation and Development (OECD)) were also included in Annex II and were expected to provide finance to support developing countries. Central and Eastern European Countries and those of the former Soviet Union were considered to be "economies in transition" and were accorded some additional flexibility. The remaining Non-Annex I Parties were considered "developing" countries. This division of labour was essential in 1992 to forge a global treaty to act on climate change.

Since 1992, the UNFCCC has achieved near universal participation, with 197 Parties. However, the Convention remains a framework instrument without enforceable targets. Recognising this weakness, in 1997 its Parties adopted the Kyoto Protocol. The Protocol entered into force in 2005 and contains legally binding commitments for developed countries to reduce their collective greenhouse gas emissions by 5% over the period of 2008–2012, compared to 1990. Individual targets were negotiated and agreed, ranging from cuts of 8% (including by the EU and its Member States) to growth caps of 10% as compared to 1990 levels.

These targets were agreed through a combination of suggestions by certain Parties as to what a common target should be (EU initially suggested −15%), and what others offered to commit to individually (the US offered −7%). The final, individualised emissions reduction targets were hammered through after closed room negotiations. For some Parties these represented significant reductions against business as usual emissions trends, for others, particularly in Eastern Europe and Russia, the targets, measured against historically high baseline levels of emissions in 1990, eventually turned out to be "surpluses" well above existing emissions levels. These targets were harmonised internationally to the extent that they constituted broad, quantitative emissions limitation or reduction targets set against a common base year[3] and within a common timeframe.

In addition to the first internationally agreed and legally binding targets and timetables, the Kyoto Protocol and the decisions taken later to implement it[4] contain detailed and rigorous reporting requirements, as well as the accounting rules and tracking systems necessary to check on Parties' compliance with their targets. Developed country Parties' targets were converted into individual carbon budgets (denominated in "Assigned Amount Units" – each Unit corresponding to a metric-tonne of CO_2-equivalent). Assigned Amount Units not used by a Party could be traded with another Party for the purpose of remaining within the latter Party's individual carbon budget. The US insisted on this arrangement much to the displeasure of the EU at the time, as countries could simply comply by buying part of the surplus of the Economies in Transition. The Protocol also established the Clean Development Mechanism, the first international mechanism for certifying carbon offsets generated by projects in developing countries that could be used by developed country Parties to remain within their budgets.

Compliance with the Kyoto targets and its carbon market rules is overseen by the Enforcement Branch of its Compliance Committee, which has the authority to suspend the right to trade and to penalise Parties for failing to remain within their budgets. As designed, under the Kyoto compliance system, a Party found to have exceeded its carbon budget, or "assigned amount," during the first commitment period of the Kyoto Protocol must deduct that excess emissions from its "assigned amount" in the subsequent commitment period at a penalty rate of 1.3.

The EU and its Member States ratified the Kyoto Protocol in April 2002. By that time, however, the US had decided not to follow up its signature of the Protocol by ratification, which was a considerable blow to the newly emerging multilateral approach. In the absence of the US, the ratification by Russia was necessary to trigger the emissions-based threshold for entry into force. This new situation required an intensive diplomatic initiative by the EU. Within weeks after the declaration by US President George W. Bush and EU Commissioner for the Environment Margot Wallström undertook a world tour to garner support for the survival of the Kyoto Protocol that brought her to Tehran,[5] Moscow, Beijing and Tokyo, as well as other cities. Several additional visits were paid to Moscow. On 18 November 2004, Russia finally submitted its ratification instrument and the Kyoto Protocol entered into force 90 days later on 16 February 2005.

The entry into force of the Kyoto Protocol was important for the world's efforts on climate action but also for Europe. Preparing for and implementing the Kyoto Protocol directly shaped the design of the EU's targets, its rules for monitoring, reporting and verifying emissions and, significantly, the EU's Emissions Trading System. These policies contributed to the EU overachieving its target of an 8% reduction below 1990 levels by three percentage points by the end of the Protocol's first commitment period in 2012.

Conclusion: The EU has consistently pursued the goal of tackling global environmental problems through UN institutions. EU climate policy has both shaped and been shaped by the Kyoto Protocol, which entered into force following considerable diplomatic efforts by the EU.

2.2 From the failure of Copenhagen (2009) to the success of the Paris Agreement (2015)

Since 1992, the world has changed considerably. Back then, it was somewhat easier to describe the world as divided between "developed" and "developing" nations, and this was reflected both in the Framework Convention on Climate Change and the Kyoto Protocol. Today, more than 25 years later, the EU, the US and Japan represent a lower share of world economic activity, due to the impressive rise of new emerging economies, most dramatically in South-East Asia. While nearly 1.1 billion people have moved out of extreme poverty since 1990, in 2013, 767 million people in the developing world still lived on less than $1.90 a day. At the same time, more than 20 Parties considered to be "developing" under the UNFCCC have *per capita* incomes higher than that of the EU's poorest Member State.

These profound economic changes were reflected, as one would expect, in the emissions pattern of the countries concerned. In Figure 2.1 the dramatic increase of China's greenhouse gas emissions is striking and represents more than a doubling in less than a decade. It starts around the years 2002/3, i.e., shortly before the entry into force of the Kyoto Protocol.

This rapidly changing context led to an intense debate on the kind of international climate change regime that should be in place after the Kyoto Protocol's first commitment period ended, in 2012. The Europeans preferred to extend the Kyoto Protocol and were prepared to continue taking the lead by signing up to a second commitment period of legally binding emissions reduction targets, with the understanding that appropriate criteria and parameters should be agreed so as to include emerging economies in taking on quantitative obligations over time. The US, always sceptical about legally binding commitments and increasingly concerned about its competitive relationship with China, was interested in a more voluntary, "bottom up" approach that would treat all countries in the same manner. While the views of developing countries were increasingly divided, the major economies within the G77 were unwilling to contemplate binding commitments under

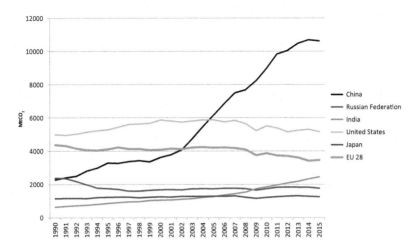

FIGURE 2.1   Global CO_2 emissions from fossil fuel use and cement production, 1990–2015

Source: EDGAR database, JRC, European Commission[6]

the Kyoto Protocol but still wanted to capture as many developed countries as possible in a second commitment period.

In 2005, at COP11 in Montreal, it was decided that negotiations on the second commitment period under the Kyoto Protocol should start. In 2007, at COP13 in Bali it was then decided that a two-track approach would be pursued. Developed countries under the Kyoto Protocol (with the likely exception of the United States and others) would by 2009 negotiate new, binding commitments under the Protocol. The other track would, within the same timeframe, negotiate an "outcome" of an undefined legal character that would represent "long-term cooperative action" and that would include the identification of "nationally appropriate mitigation actions" for all countries.

This vague, lopsided mandate and the tensions between the different groups of countries came to a head at COP15 in Copenhagen in 2009. By then it was clear that despite its strong rules and innovative market and compliance mechanisms, the Kyoto Protocol and its exclusive focus on developed country targets would not prove to be a long-term model for a more inclusive future UN climate regime. No alternative model for a legally binding agreement to replace the Kyoto Protocol gained consensus. In its place, the US vision for a more "bottom up," less binding approach received the support of the emerging economies and the Europeans found themselves isolated, despite huge support from the NGO community. In the final days of the COP, the formal negotiating tracks broke down and were overtaken by an *ad hoc* group of 28 government leaders that produced the "Copenhagen Accord."

The Accord proposed a system of "pledge and review" whereby all Annex I (developed) Parties would commit to implement individually or jointly the quantified economy wide emissions targets for 2020, whereas Non-Annex I (developing country) Parties to the Convention would implement "mitigation actions." These pledges of developed and developing countries would be compiled by the UNFCCC Secretariat in separate documents and be subject to distinct review processes. Under the Accord, developed countries would commit to a goal of jointly mobilising US$100 billion dollars a year by 2020 to address the needs of developing countries, and that a significant portion of such funding should flow through a newly established Green Climate Fund. The unorthodox way in which the Accord was negotiated and presented led to opposition from many countries and the COP ended in confusion and acrimony.

Nevertheless, in order to rescue the situation, by the end of January 2010 more than 90 countries, including the EU, its Member States and all major economies, had submitted voluntary 2020 emission reduction pledges. This allowed negotiations to get back on track at COP16 in Cancun. In fact, Parties agreed to the essential elements of the Copenhagen Accord, namely a common but differentiated system of "pledge and review," which called on all Parties to participate in emission reductions but retained the "bifurcated," or differentiated, categories of Convention Annexes. The pledges were differentiated between the "targets" of Annex I countries and "actions" by non-Annex I countries and were to be measured, reviewed and verified by two similar but differentiated systems. This, along with the US$100 billion pledge from developed countries, became the backbone of the international climate regime for the eight years between 2013–2020.

In this context, to demonstrate both a willingness to lead and support for strong multilateral rules, the EU and its Member States agreed to take on a new and ambitious set of targets under the "Doha Amendment." This was possible in large part because, by 2012, the EU had already put in place all the key elements of its regional and national climate policies up to 2020 in the course of implementing the Protocol's first commitment period. The EU's Doha Amendment target is therefore a reflection of the targets set regionally and nationally through the evolution of what had become EU climate policy. Meanwhile, Canada formally withdrew from the Kyoto Protocol while Japan, Russia and a number of other industrialised countries declared their intention not to enter into a second commitment period under the Kyoto

Protocol. As of early 2019, the Doha Amendment has not yet met the threshold of ratifications necessary to bring it into force.[7] As the end of the Doha Amendment's commitment period approaches in 2020 and the Paris Agreement becomes operational, the Kyoto Protocol itself will become irrelevant.

For the period after Copenhagen, the EU and its closest negotiating partners, including an increasingly engaged US Administration under President Barack Obama, recognised that neither an extension of the Kyoto Protocol nor a purely voluntary system of pledge and review that continued to differentiate responsibilities between developed and emerging economies was a sufficient response to the urgency of climate change. In 2011, at COP17 in Durban, the EU together with the most vulnerable developing countries led a "progressive alliance" of developed and developing countries to win acceptance of a new mandate to negotiate "a Protocol, another legal instrument or an agreed outcome with legal force under the Convention applicable to all Parties, which is to be completed no later than 2015."

In November 2015, 23 years after the adoption of the Convention in 1992, the text of the Paris Agreement was adopted by "acclamation."[8] The EU and a progressive alliance of developed and developing countries, now together with the US, supported an ambitious text put forward by the French Presidency of the Conference. Drawing on lessons from experience of the UNFCCC and the Kyoto Protocol, the Paris Agreement combines ambitious, science-driven goals, nationally determined emissions reduction contributions by all Parties and robust systems of transparency and accountability that is applicable to all Parties. While it provides important flexibilities for developing countries, these are provided for on the grounds of constraints in their national capacities rather than their categorisation under the UNFCCC. Developed countries, which are also not defined under the Paris Agreement, are expected to continue to provide support to developing countries both for cutting emissions and preparing for the impacts of climate change, including by continuing to follow through on the US$100 billion a year collective pledge made in Copenhagen beyond 2020.

More than 150 Heads of State and Government attended the Paris Conference to express their support for global action on climate change. When the Agreement was opened for signature six months later at the United Nations Headquarters in New York on 22 April 2016, it was signed by 174 countries and by the European Union. It entered into force less than a year after it was adopted, on 4 November 2016, when 55 Parties to the Convention representing more than 55% of global emissions had deposited their instruments of ratification. At the time of writing it has 184 Parties.

President Trump's announcement, in June 2017, of his intention to with-draw from the Paris Agreement was a great disappointment. That was a second significant attack by the US on the multilateral approach to climate change, following President Bush's rejection of the Kyoto Protocol in 2001. The announcement was received coldly, but rather than derailing the Paris Agree-ment this action has led to the international isolation of the US Administra-tion with regard to climate diplomacy and the mobilisation within the US of more climate action by non-federal government actors. It even prompted the two remaining countries not yet to have signed the Agreement, Syria and Nicaragua, to do so. The earliest that a country can leave the Paris Agree-ment is 4 November 2020, and the earliest the US can give official notice of its intention to leave is one year before that, on 4 November 2019.

Conclusion: The Paris Agreement is the new multilateral structure for global action on climate change. It requires action by all countries and is currently endorsed by the whole world except the US.

2.3 Essential features of the Paris Agreement

The Paris Agreement is one of the first of a new generation of multilateral environmental agreements in that it combines a number of "top-down" and "bottom-up" elements. In this way, it allows for much more differentiation of policy action by its Parties.

2.3.1 Applicable to all Parties

The speed and near universality of the Paris Agreement ratifications demon-strates broad political support for its essential characteristics: it sets ambitious collective goals, allows each Party to determine its own targets and timetables and demands transparency and accountability from its Parties. It is the first international climate agreement that is "applicable to all" Parties. It provides flexibility for developing countries on the basis of differences in national capacity rather than the defined categories set under the UNFCCC and perpetuated under the Kyoto Protocol.

From 2020 onwards, the Paris Agreement will in effect replace the twin track approach of the Kyoto Protocol and voluntary pledges. It is a landmark agreement that will continue to have effect for decades to come. In princi-ple, all Parties are now being dealt with in the same manner as regards their obligations on emissions reductions, but these reductions are defined through plans ("Nationally Determined Contributions" or "NDCs") that each Party draws up for itself. The flexibility provided through this "bottom-up" element is an expression of the UNFCCC core principle of "common but differentiated responsibilities and respective capabilities," taking account of each Party's national circumstances and capacities.

While for Europeans the Paris Agreement is a firm expression of the multilateral approach they strongly prefer, a significant concession is made in accepting the major bottom-up element of Parties determining their own contributions and level of ambition. However, the Paris Agreement will allow for strong peer pressure on Parties through increased transparency and regular reviews.

The Paris "rulebook," adopted in 2018 at COP24 in Katowice, Poland, set out detailed guidance on how the Agreement's transparency and accountability framework and ambition cycle will operate. The outcome respected the careful balance struck in Paris, but by providing greater clarity and detail will enhance the "top-down" nature of the regime.

2.3.2 Ambitious collective goals

Building on the guidance offered by the International Panel on Climate Change (IPCC) and the Convention's objective of limiting concentrations of greenhouse gases in the atmosphere to levels that would prevent dangerous climate change, the Paris Agreement clarifies that global average temperature rises as compared to pre-industrial levels must stay "well below 2°C" while "pursuing efforts to limit such a rise to 1.5°C." These "temperature goals" help to define what the international community considers to be dangerous climate change and set an overall ambitious direction for the development of Parties' individual and collective efforts.

The Paris Agreement also wants to achieve a balance between sources and sinks of emissions in the second half of this century. In other words, it describes as its purpose a profound and global transformation over the next decades from an economy primarily dependent on fossil fuels to one that has reached a steady state in which global emissions are at "net zero" and atmospheric concentrations of greenhouse gases have balanced out at levels consistent with the temperature goals.

The Paris Agreement goals address global emissions and thus have the potential to cover all sources of emissions that contribute to anthropogenic climate change, including those originating from international aviation and maritime operations.[9] Both international transport sectors show rapidly increasing emissions and these will have to be addressed respectively in the context of the International Civil Aviation Organisation (ICAO) and the International Maritime Organisation (IMO) if the ambitious goals of the Paris Agreement are to be achieved.

2.3.3 Dynamic, five-year ambition cycles

Under the Paris Agreement, each Party commits to "prepare, communicate and maintain successive Nationally Determined Contributions that it intends to achieve"[10] every five years. Each successive contribution will represent a progression over the previous one and shall be informed by a global stock take of Parties' collective progress towards the Agreement's long-term goals. In 2018, an initial Facilitative "Talanoa" Dialogue was held, taking into account the results of the IPCC Special Report on 1.5°C. It also prepared for 2020 when Parties with a 2025 commitment, such as the US,[11] Brazil and South Africa are expected to communicate their post-2025 targets.

Formally, the first of these five-year ambition cycles will begin with a global stock take in 2023, with an expectation that Parties will communicate their post-2030 targets by 2025. This and future cycles will be essential to closing the gap between current and announced emissions targets and contributions. These reviews will also be necessary to reach the Paris Agreement's goals of limiting temperature rise, peaking global emissions and achieving "net zero" emissions.

Although the ambition, form and content of Parties' targets and contributions will remain nationally determined, the Paris Agreement puts in place rules and processes that will encourage their harmonisation, quantification and comparability over time. Parties agreed to continue negotiations on common features that will be applicable to future rounds of targets. Developed country Parties are expected to have and maintain the most robust form of target, namely economy-wide absolute emissions reduction targets like the EU's. Developing countries are expected to move towards economy-wide emissions reduction or limitation targets over time, and future rounds of targets and contributions will be subject to common accounting rules that may also be applied to the first round of targets on a voluntary basis.

2.3.4 Transparency and accountability

The Paris Agreement establishes a robust, legally binding transparency and accountability framework that is applicable to all Parties. Together with the "rulebook," the Agreement sets out rules, institutions and procedures for the measurement, reporting and verification of information provided by Parties through national inventories of emissions and the policies they have put in place to achieve their targets. This will enable the tracking of progress of each Party towards its target, as well as an understanding of collective progress towards the Agreement's goals. The previous split approach between developed and developing countries operating under the Convention and the Kyoto Protocol, which required very little of Parties classified as developing countries, will be phased out after the submission of reports regarding data for the year 2020.

The transparency framework makes it clear that all Parties must report, at least bi-annually, greenhouse gas inventories and information necessary to track progress with the mitigation contributions in accordance with agreed methodologies and common metrics. Only the Least Developed Countries (LDCs) and the Small Island Developing States (SIDS) enjoy flexibility with regard to the frequency of reporting.

The Agreement also includes an obligation on each Party to account for anthropogenic emissions and removals relating to their targets in a way that promotes environmental integrity, transparency, accuracy, completeness, comparability and consistency and to ensure that any double counting arising from the use of carbon markets is avoided. These common rules, known as the "rulebook," are essential to ensure that targets are implemented and to promote trust in the international process. Each Party's report shall undergo a technical expert review, and each Party shall participate in a facilitative multilateral consideration of its performance.

The rulebook elaborates on how the transparency system will provide flexibility for those developing countries that need it in light of their capacity. These flexibilities were negotiated on a case-by-case basis to allow, for example, developing countries to report their national inventories less frequently or with regard to fewer greenhouse gases. These countries must concisely clarify the capacity constraints they are facing and indicate estimated timeframes for overcoming these constraints.

The transparency system will be supported by a Committee on Implementation and Compliance, designed to both help and hold accountable countries experiencing challenges with the implementation of and compliance with the mandatory provisions of the Agreement and the rulebook. While this committee is facilitative, non-adversarial and non-punitive in nature, it can engage individual Parties regarding their performance and provide advice, recommendations to the Agreement's finance institutions, assist in the development of implementation plans and in certain circumstances issue findings of fact. This will bring public and political attention to the challenge of implementation.

The EU's new Energy Union Governance Regulation[12] meets the require-ments of this transparency and accountability framework and a number of features were updated in light of the Paris outcome, namely the alignment with the overall Paris ambition cycle. While a work programme has been launched to develop common accounting rules, including for land, these will not apply to Parties' first mitigation target under the Paris Agreement. The EU will work closely with other Parties to ensure any internationally agreed approaches including the accounting for emissions from land are consistent with EU approaches.

The framework will also provide for the transparency of the Agreement's provisions on adaptation and on climate finance, capacity building and tech-nology transfer, as discussed later.

2.3.5 Increasing resilience to and responding to the adverse effects of climate change

The Paris Agreement establishes, for the first time, a global goal on adapta-tion with the aim to enhance capacity, climate resilience and reduce climate vulnerability. Internationally, it encourages greater cooperation amongst Par-ties to share scientific knowledge on adaptation as well as information on practices and policies. As part of this international cooperation, developed country Parties must also continue to provide, as part of their commitments on climate finance, resources to developing country Parties to support their adaptation efforts.

All Parties' efforts to promote adaptation must "represent a progression over time" and the five-year global stocktaking described earlier as part of the Agreement's "ambition cycle" additionally applies to adaptation. It will review overall progress in achieving the Agreement's adaptation goal includ-ing by reviewing the adequacy and effectiveness of the support provided for adaptation.

The Paris Agreement acknowledges that addressing "loss and damage" resulting from climate change is a specific aspect of increasing resilience to the adverse effects of climate change. Many vulnerable developing countries, especially low-lying and Small Island Developing States, are struggling with how to prepare for and manage loss and damage associated with extreme weather and the slow onset impacts associated with climate change. Never-theless, the decisions taken in Paris clarify that the Paris Agreement provi-sions on loss and damage do not involve or provide a basis for any liability or compensation.

2.3.6 Fostering cooperation and financial flows

The Paris Agreement also fosters cooperation amongst Parties by encouraging the responsible use of international carbon markets and the mobilisation of support to developing countries. Implementing the emissions targets will require very substantial policy action and investments in clean technologies in the coming years in all countries. The Paris Agreement includes the aim of "making financial flows consistent with a pathway towards low greenhouse gas emissions and climate resilient development." Shifting and rapidly scaling up private investment is essential to the transition to a low-emission and climate resilient economy and to avoid "locking-in" high emission infrastructure.

In Paris, the EU, its Member States and other developed country Parties committed to continuing, in the period from 2020 until 2025, the goal set in Copenhagen to mobilise US$100 billion annually from public and private sources by developing countries. Before 2025, the Parties to the Paris Agreement will set a new collective quantified goal from a floor of US$100 billion per year. This will provide an opportunity to broaden the donor base to include countries previously considered only as recipients of assistance.

> Conclusion: The Paris Agreement is applicable to all Parties in a similar fashion: each Party determines its own target or contribution, and all Parties are ultimately subject to a common, transparent governance system. Flexibilities are provided for those developing countries that need it but based on gaps in their capacity.

2.4 Are global emissions slowing down?

The Paris Agreement is ambitious in its objectives and calls on Parties to reach global peaking "as soon as possible" and to "undertake rapid reductions thereafter." It assumes that global emissions have not yet reached the maximum level and will keep increasing year on year. Since Paris, emissions first stabilised and then went up again in 2017. However, it is encouraging that in many countries, emissions have either gone down over time in absolute terms or the rate of increase is slowing down and that a gradual decoupling from economic growth is happening.

Developing countries in general, but emerging economies in particular are allowed to increase their emissions temporarily under the Paris Agreement.

The fact is that their economic development leads to an increase in emissions. However, they are also well aware that they are contributing to climate change that will be experienced in the future and are therefore willing to invest in low-carbon technology. China announced in its NDC that its emissions would peak no later than in 2030. This means that its historical emissions – although much more recent than those of UNFCCC Annex 1 countries – will continue to rapidly accumulate. Recent analysis seems to indicate that that the date of peaking of China's emissions could be before 2025.[13]

Before coming to Paris, Parties were asked to submit their plans for their NDCs in order to confirm their good faith and show their readiness to act within the new bottom-up approach. Several analyses have been made since then trying to summarise the overall result of the pledges and policy intentions, despite the many different definitions of targets and pathways that were advanced. The Climate Action Tracker regularly assesses projections on the basis of existing pledges and policy declaration. The most recent of these shows that without the NDCs or any other policy effort, the

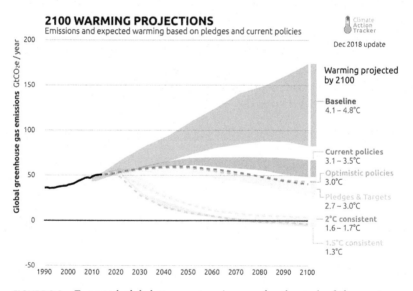

FIGURE 2.2   Expected global temperature increase by the end of the century compared to pre-industrial levels implied by global emissions pathways in six scenarios[14]

Source: Climate Action Tracker Warming Projections Global Update, December 2018[15]

world would enter a scenario of global warming ranging between 4.1° and 4.8°C. If fully implemented, existing policy declarations, including those of NDCs, would bring this down to a range of 2.7°-3.0°C. In short, the 2°C goal of the Paris Agreement is not yet within reach, and a significant tightening of the NDCs will be required when new pledges are due in 2020 and 2025.

While the overall result of the first round of NDCs is insufficient, the exercise of the NDCs remains most valuable. It requires all Parties to go into a systematic screening of the emissions impact of their economy. Moreover, all Parties are different. For some, as is the case for the EU, the emphasis is on carbon emissions from industry and power, while in some others, such as New Zealand, the bulk of emissions come from agriculture. In the case of Brazil or the Democratic Republic of Congo, most emissions are linked to tropical deforestation.

The NDC policy plans also allow countries to reflect on their emissions per person, where there are major differences. As already illustrated in Figure 1.5 in the previous chapter, in 2015, CO_2 emissions *per capita* were 6.9 tonnes CO_2 in the EU and have shown a steady reduction since 1990. Chinese emissions reached EU levels in 2012 and have since then continued growing to reach 7.7 tonnes CO_2 *per capita*. Although falling slowly, the *per capita* emissions of the US (at 16 tonnes *per capita* in 2015) are very high compared to the global average and are more than twice those of the EU. Levels *per capita* in India at 1.85 tonnes in 2015, for example, remain low but are steadily increasing.

Finally, the NDCs also include valuable information for policy research. While some NDCs are clear, transparent and comprehensive, others require improvement and the exchange of "good practice" in policy making terms has considerable potential. Taking account of the commitments of the Paris Agreement, the IEA has looked into the future to see the amount of worldwide energy investments planned each year in the period from 2018–2040 (see Figure 2.3). It is estimated that, on the basis of policies and targets announced by governments, investment during the period from 2026–2040 will amount to US$2821 billion, of which US$1559 billion, representing over 55%, will be invested in renewable energy, transmission and distribution, energy efficiency and biofuels. Even then, the IEA warns, "While the picture brightens, there is still no peak in global energy-related CO_2 emissions." This analysis, combined with other information from the NDCs, provides useful insights into future investment opportunities and where these investments will be located.

World	New policies scenario	
Yearly average investment (billion, $2017)	**2018–25**	**2026–40**
Fossil fuel supply	852	992
Biofuels supply	9	18
Power sector	**810**	**899**
Coal	56	43
Gas	55	46
Oil	4	2
Nuclear	51	45
Utility-scale batteries	9	15
Renewables	322	361
Hydro	70	75
Wind	98	119
Solar PV	127	116
Transmission and distribution	313	387
Total final consumption	**545**	**912**
Efficiency	397	666
Renewables	116	127
Other	32	119

FIGURE 2.3 Global yearly average investment (billion US$, 2017)

Source: International Energy Agency (2018), World Energy Outlook 2018, OECD/IEA, Paris

> Conclusion: The NDC instrument is a most useful tool for policy making. NDC policy plans submitted under the Paris Agreement are not yet aligned with delivering the 2°C goal. They first need swift and solid implementation and then an ambitious review, which started with the Facilitative Dialogue in 2018 and leads to the global stock take in 2023.

2.5 EU's international cooperation focuses on the implementation of low-carbon policies

2.5.1 Focus on implementation in major economies and developing countries

The EU plans to increase its efforts in sharing its own experiences on designing and implementing climate and energy policies, in particular with other major economies. This includes countries like China and South Korea that are setting up emissions trading systems as well as a broader range of countries, including all major economies that are deploying renewable energy technologies, improving their energy efficiency policies and their clean mobility policies. Harmonising reporting on efficiency standards amongst G20 countries is another means of encouraging economies of scale and the lowering of technology costs. Supporting densely forested developing countries to reduce the emissions from deforestation and forest degradation, particularly when it comes to the monitoring through its COPERNICUS earth observation programme from space, will continue to play a role in EU development cooperation. Where the EU has established special relationships with "candidate countries" (that have applied to join the European Union) and in the EU neighbourhood region, particular emphasis will be placed on practical policy dialogue and technical assistance.

Many developing countries are, for the first time, committing to reducing or limiting their emissions and/or to increasing their efforts to adapt to climate change. In addition, they have subscribed to implementing a new transparency regime. The Paris Agreement recognises that these efforts need to be supported, especially in those developing countries that are most vulnerable to climate change and/or that lack capacity. In this context the synergies between climate action, financing for development under the Addis Ababa Action Agenda and the Agenda 2030 associated with the Sustainable Development Goals need to be fully exploited.

The close coordination of trade rules and climate policies, wherever possible, will be increasingly important to align international markets, national standards and climate-related policies. For example, EU policies on renewable energy are designed to ensure that domestic and imported biofuels and biomass are sustainably produced if they are to count towards EU emissions reduction efforts. EU bilateral and regional trade agreements increasingly include commitments from Parties to fully implement the Paris Agreement, and not to lower their climate ambitions as a means of attracting trade or investment.

2.5.2 Mobilising finance domestically and internationally

Mobilising the necessary finance for the transition to low-carbon economies will be a major challenge for all, both developed and developing countries.

For the EU to achieve the targets of the Paris Agreement, including the "at least 40%" cut in greenhouse gas emissions by the EU in 2030, around €180 billion of additional investments a year are needed. The scale of this investment challenge is beyond the capacity of the public sector alone. The financial sector has a key role to play in reaching those goals, as large amounts of private capital should be redirected towards sustainable investments. Various domestic initiatives have been taken in the EU in order to intensify private finance in the coming years that could serve as an example internationally:

- A key priority of the EU's Capital Markets Union's (CMU) is the fostering of more sustainable private investments. In 2018, the Commission launched an Action Plan on Financing Sustainable Growth containing a roadmap and seeking a lead on global work in this area, with legal proposals.[16]
- Public sector finance should leverage private finance into strategic areas. In this respect, the EU extended and reinforced the European Fund for Strategic Investments (EFSI 2.0).
- For the period between 2021–2027, the European Commission has proposed that the EU increases its target from 20% to 25% of its entire budget spending to be climate relevant.[17] This would further mainstream climate issues into funding for regional development, research and innovation, the common agricultural policy and international cooperation.
- Revenues from the auctioning of more than 800 million emission allowances under the EU Emissions Trading System between 2021 and 2030 will be directed into an Innovation Fund and a Modernisation Fund. At a price of €20 per tonne, this would represent some €16 billion to assist in funding renewable energy, energy efficiency and industrial innovation.

Internationally, the EU is also supporting discussions that aim to ensure an effective and coherent approach to green finance globally. At the international level, the Green Climate Fund, international financial institutions such as the World Bank, the European Investment Bank, the European Bank for Reconstruction and Development and national development banks active in developing countries increasingly engage in the implementation of the Paris Agreement.

Since Paris, the EU and its Member States have continued to contribute the lion's share of global public climate finance. In 2016 and 2017, the EU's contribution rose to more than €20 billion, a significant increase compared to earlier years.

Conclusion: The EU is shifting its international cooperation and financial support, both public and private, towards implementation of the Paris Agreement.

Conclusion

The Paris Agreement that entered into force in 2016 is the new solid base for the global effort on climate change for the coming years. It has secured global participation in record time. The Paris Agreement has made the UN Framework Convention on Climate Change operational and more comprehensive. This is a major achievement given the political context, in which multilateral efforts are coming under increasing pressure.

One of the main characteristics of the Paris Agreement is that the distinction between "developed" and "developing" countries is far more nuanced than under the Convention and the Kyoto Protocol, placing greater emphasis on national circumstances. The Convention principle of "common but differentiated responsibilities and respective capabilities" has been secured through the bottom-up approach of the Agreement. The policy plans (NDCs) that each Party submitted, and which have been made official by virtue of the ratification process do not yet add up to meet the "well below 2°C" objective. Achieving this will require a tight management of the five-year ambition cycle. Another main challenge is to ensure that all Parties follow the detailed rules on transparency and accountability agreed at COP24. This will require a combination of capacity building support, technical expert review and peer pressure to create the right combination of incentives for implementation and compliance.

The Paris Agreement has also some weaknesses. It does not foresee negotiation around the Nationally Determined Contributions according to an agreed set of precise parameters. There is also no strong guidance from the multilateral level on which policies need to be implemented nationally. Similarly, the compliance regime does not foresee sanctions but is based on peer pressure. As a consequence, all key implementation issues are left to the individual Parties. That means a great deal of political effort will have to be put into difficult implementation issues in every single country for years to come. It will require a tremendous amount of political goodwill, internationally but even more so at the national level, within individual governments and with their respective stakeholders.

In addition, the Paris Agreement does not explicitly "organise" the development of common policies or standards, as the Kyoto Protocol did, for example, with respect to international carbon markets.[18] It is therefore key that groups of like-minded countries come together to foster common policy plans, to share their experiences and knowledge of policy making to facilitate implementation. There is much scope for action, for example by international financial institutions such as the World Bank and its regional sisters, to foster such cooperation.

In this respect, the EU has already agreed legally binding targets for its 28 Member States until 2030. These targets are differentiated according to their relative prosperity and, based on economic analysis, incorporate the principle of cost-efficiency. The EU also developed common policy instruments such as the EU Emissions Trading System in the field of renewable energy, energy efficiency and CO_2 standards for cars and appliances. At the same time, a considerable effort is being undertaken to mainstream climate action into other policies such as research, industry, finance, trade, agriculture and development cooperation.

The EU is ready to deploy extensive efforts in the field of international cooperation and financial support both with emerging economies and developing countries for the solid implementation of the Paris Agreement. The involvement of all stakeholders from the private sector, NGOs and local authorities will be key to a successful outcome. The blunt attack of US President Trump on the Paris Agreement has so far not shaken support for the Agreement within the EU and around the world. Fortunately, the economics of the low-carbon economy are improving rapidly with clean and sustainable technologies becoming cheaper by the day, and momentum is building across the globe, meaning that the actions of one country ought not to destabilise the efforts of rest of the world.

It must be recognised that recently the number of global problems confronting politicians has dramatically increased: wars and political instability causing an upsurge of poverty and refugees, respect for democratic institutions and for expert evidence and the continued undermining of rules-based multilateral institutions, even established ones such as the WTO. This leaves climate change as only one of many global challenges to deal with. Today's politicians may mistakenly think that their work on climate change is over now that the Paris Agreement has entered into force. The sobering reality is that all implementation issues relating to the Nationally Determined Contributions remain squarely on the table and require increasing political attention given the seriousness and the urgency of the climate problem.

Notes

1 The Club of Rome, founded in 1968, is an organisation of individuals who share a common concern for the future of humanity and strive to make a difference. Members include notable scientists, economists, businessmen and businesswomen, high-level civil servants and former Heads of State from around the world.
2 The UNFCCC was adopted on 9 May 1992 and opened for signature at the United Nations Conference on Environment and Development that took place from 3 to 14 June 1992.

3 Given their particular circumstances, "economies in transition" were allowed to choose one among a limited number of base-years.

4 These decisions were mostly taken in the Marrakech Accords at the 7th Conference of Parties (COP-7) in 2001.

5 Iran being the chair of the G77 and China grouping at the time.

6 Janssens-Maenhout et al. (2017).

7 As of 21 February 2019, 126 Parties have ratified the Doha Amendment, whereas 144 ratifications are needed for it to come into effect. https://unfccc.int/process/the-kyoto-protocol/the-doha-amendment.

8 "By acclamation" means with no state objecting in the plenary session of the Conference of Parties.

9 Outgoing aviation emissions are included in the EU 2020 targets, and in the EU 2030 legislation and countries can decide to include these in their Nationally Determined Contributions (NDCs) under the Paris Agreement.

10 Article 4 of the Paris Agreement.

11 However, the US has announced its intention to withdraw from the Paris Agreement.

12 Regulation (EU) 2018/1999 of the European Parliament and of the Council of 11 December 2018 on the Governance of the Energy Union and Climate Action. Official Journal L328, pp. 1–77 published on 21.12.2018.

13 See: "China's "new normal": structural change, better growth, and peak emissions", by Fergus Green and Nicholas Stern, Policy brief by the Centre for Climate Change Economics and Policy (CCCEP) and the Grantham Research Institute on Climate Change and the Environment, June 2015. www.lse.ac.uk/GranthamInstitute/publication/chinas-new-normal-structural-change-better-growth-and-peak-emissions/.

14 Baseline emissions, emissions compatible with warming of 1.5°C and 2°C, respectively, and the three scenarios resulting from aggregation 32 country assessments: pledges and targets, current policies and an optimistic scenario. Ranges indicated uncertainty in emissions projections; dotted lines indicate median (50%) levels within these ranges. For further details. https://climateanalytics.org/media/cat_temp_upadate_dec2018.pdf.

15 The Climate Action Tracker (CAT) is an independent scientific analysis produced by three research organisations tracking climate action since 2009: Climate Analytics, Ecofys and NewClimate Institute.

16 Legal proposals have been made to regulate benchmarks for low-carbon investment strategies (provisionally agreed), a proposal to establish a unified EU classification system ("taxonomy") of sustainable economic activities and a proposal to improve disclosure requirements related to sustainability risks and opportunities. See http://europa.eu/rapid/press-release_IP-19-1418_en.htm.

17 The Multi-annual Financial Framework 2021 to 2027 is still being negotiated with the Member States and European Parliament.

18 However, the rules relating to the implementation of Article 6 of the Paris Agreement will include rules relating to carbon markets.

3

HOW ECONOMIC ANALYSIS SHAPED EU 2020 AND 2030 TARGET SETTING

Tom Van Ierland and Stefaan Vergote

Introduction

The preparatory economic analysis for setting climate and energy targets has become increasingly sophisticated over the years. Numerous think tanks, universities and foundations from all over the world have become involved in this debate. Research teams from the European Commission and several university institutes have begun the laborious work of integrating economic, energy and climate models both for Europe and for the world.

3.1 The potential of integrated economic and climate modelling

For the Kyoto Protocol's first commitment period (2008–2012), the target setting was subject to less economic analysis than has been the case for the 2020 and 2030 targets. Kyoto's target of an 8% reduction targets for European countries was more the outcome of the international negotiations. Prior to Kyoto, the then 15 EU Member States had declared their willingness to assume a 15% reduction target compared to 1990, even though the Member States were unable to agree on a "burden sharing" of all of this 15% in advance. This rather detracted from the credibility of the proposal by making it look as if the EU did not really expect such a figure to be agreed. The 15% target proposed by the Europeans was conditional. At the time, Europeans wanted minimum differentiation between the Kyoto targets, and some even hoped that all developed countries would adopt the same target.

Some Member States did some economic modelling at the time, in particular the Netherlands and the UK, but the European Commission did not take a leading role in fixing these targets for the Kyoto Protocol's first commitment period. The most important issue at European level was the "bubble" arrangement of Article 4 of the Kyoto Protocol, which allowed EU Member States to differentiate targets between themselves on the condition that the EU collectively met the 8% reduction target that it finally assumed in Kyoto. Such differentiation was negotiated within the Council in 1998, without a specific proposal from the Commission and without the involvement of the European Parliament.

By contrast, the preparation of the targets for the period from 2012–2020 was subject to much more extensive economic analysis. The Commission took the lead and was relied upon by the Member States as a whole. The increase in the number of Member States by 2008 from 15 to 27 made the task much more complicated. Furthermore, in the context of 2020 targets, the EU had to prepare for a political scenario in which it had to continue its climate action policies without any guarantee that the rest of the world would follow. Under the Kyoto Protocol, the emerging economies had no obligations to reduce their emissions, and in 2001, the US disengaged altogether. The second commitment period of the Kyoto Protocol – called the "Doha Amendment" – was only agreed to in 2012, with several large industrialised countries not supporting a second commitment period – for example Canada, Japan, Russia and the US.

The fact that the Europeans were formulating a commitment that might not be matched by others made it imperative for Europe to proceed carefully. Thorough economic analysis for 2020 focused not only on the formulation of a unilateral target for the EU but was also accompanied by a more ambitious target in case the rest of the world would follow. The result was that the EU eventually agreed to a 20% unilateral greenhouse gas reduction target for 2020 compared to 1990, but declared its willingness to increase its ambition to 30%, "provided that other developed countries commit themselves to comparable emission reductions and economically more advanced developing countries to contributing adequately according to their responsibilities and respective capabilities."[1] What Europe was willing to do depended on what others committed to undertake. Subsequently, as the international negotiations endeavoured to include most countries of the world in any future agreement, in particular the US and the emerging economies, it became clear that a much wider differentiation of effort would have to be allowed internationally to take account of the wide variety of circumstances across the globe.

It is in this context that the analysis of the EU targets for the phase 2020–2030 was undertaken. A raft of policy scenarios was compared, each of them accompanied by quantitative estimates of the climate result, the overall economic impact, the expected sectoral developments and the associated economic costs and benefits – including in terms of competitiveness. It is important to note that it was never the intention to predict or forecast the future. What these quantitative analyses did, however, was to give insights on what would be technically and economically realistic within a set time frame, analyse interactions between different policy instruments and provide an outline of the key political parameters.

The economic analysis allowed for the formulation of a reliable order of magnitude of the overall economic impact, of the likely technologies that would have to be brought to the market, and on the accompanying measures that would have to be considered. In this context, the modelling was also capable of providing a good insight into the distributional issues that the transition would bring about, both in relative shifts between economic sectors but also between different regions and Member States within the EU. In sum, it provided a solid analytical base for proposals that imply substantial economic transformation, through which it additionally created a negotiation space between Member States and with the European Parliament.

Conclusion: Contrary to the approach followed by the EU for the first commitment period under the Kyoto Protocol, extensive modelling showed the likely economic impact of different climate targets for 2020. It allowed for informed choices to ensure both fairness and cost-effectiveness.

3.2 The Low-Carbon Roadmap towards 2050

After the considerable disappointment of the Copenhagen COP in 2009, the EU realised that the rest of the world would not join a Kyoto Protocol-type of agreement and that a review of its unilateral target for 2020 would be politically difficult. The EU had repeatedly confirmed its full commitment to the "below 2°C objective," but many pertinent questions remained on the reduction pathways and on the technological, behavioural and energy and transport-system changes that such a major transition to a low-carbon economy would imply over time. The plan was therefore to focus more on these strategic questions and to look beyond 2020. In 2011, the European Commission produced a Low-Carbon Roadmap,[2] accompanied by an Energy Roadmap[3] to flesh out the perspective through to 2050.

3.2.1 Domestic emission reductions of at least 80% by 2050

The first conclusion of the analysis was that, as part of a global effort to meet 2°C, it would be necessary for the EU to achieve domestic[4] emission reductions of at least 80% compared to 1990 in 2050. By focusing on the *domestic* effort consistent with 2°C,[5] the Roadmap concentrated on the technological, energy and economic transition required. The analysis demonstrated that international offsets can at best fulfil only a temporary role in the short- and medium-term by spreading low carbon technologies quicker to more countries and keeping transition costs manageable.

What is more, the modelling projections showed that such reductions within the EU were both technologically possible and economically feasible. The Roadmap therefore successfully focused the debate on the required domestic low-carbon transition in energy, transport and industry.

A wide range of "what if" scenarios were assessed, with different assumptions regarding technological advances and different fossil fuel prices. The analysis showed that a sufficiently strong carbon price needed to be applied across all sectors to trigger the required shift from carbon-intensive to low-carbon investments and the changes needed in energy use. An 80% reduction in greenhouse gas emissions would be possible through the large-scale deployment of innovative, existing technologies, without the need to rely on non-proven "break-through" technologies, such as nuclear fusion, or without major lifestyle changes for example dietary changes, strong upheavals in mobility patterns or the development of a hydrogen-based economy. Such promising developments could further facilitate a low-carbon economy but were not included in the analysis, given the uncertainties around their technical, societal and economic feasibility and because of the difficulties of including them in the modelling tools.

3.2.2 Milestones

The second major conclusion of the economic analysis underlying the Roadmap was that it advanced an indicative time frame for the required emissions reduction. It refrained from using the word "targets," as this word was reserved for the political debate that it was supposed to inform. Instead, the Roadmap suggested "milestones," calling for interim reductions such as 25% by 2020, 40% by 2030 and 60% by 2040, all compared to 1990 levels. The analysis showed that these milestones were the minimum reductions to be achieved over this period if the long-term target was to be realised in a cost-effective way. The milestones created a good understanding amongst

TABLE 3.1 Average yearly total investments and fuel expenses[1]

Total average yearly investments (B€)	2011–20	2021–30	2031–40	2041–50	Average
Reference	816	916	969	1014	929
Effective technologies	863	1040	1299	1589	1198
Delayed Action	845	1011	1392	1689	1234
Total average yearly fuel expenses (B€)	2011–20	2021–30	2031–40	2041–50	Average
Reference	930	1170	1259	1376	1184
Effective technologies	911	1067	1034	1019	1008
Delayed Action	922	1118	1061	993	1023

Source: PRIMES results, Impact Assessment accompanying the European Commission's 2050 Low-Carbon Roadmap; SEC(2011) 289 final of 08.03.2011

1 SEC(2011) 289 final of 8.3.2011 Impact Assessment accompanying the 2050 Low-Carbon Roadmap COM(2011) 112 final; the figures include all energy and transport related investments in all sectors (supply, demand and network).

policy makers about the level of ambition that would at least be required to respect the "below 2°C" temperature goal. They influenced and shaped the future policy discussions.

The milestone analysis also demonstrated that a so-called "delayed action" scenario, in which lower ambition in 2030 (i.e., lower than 40%) was achieved but that eventual reductions would be made that were still consistent with 2°C, was substantially more expensive than a scenario representing a cost-effective achievement of long-term targets. Table 3.1 indicates that "delayed action" until after 2030 would lower investment needs in the shorter-term but would increase overall investment costs and fuel expenses. In this way, the Roadmap prepared the ground for the emerging political debate on the 2030 target and provided further analytical grounds to aim for a greenhouse gas reduction target of at least − 40% in 2030 compared to 1990.

3.2.3 Sectoral analysis

The third outcome of the Roadmap was a detailed sectoral analysis. This elaborated how the main economic sectors of power generation, industry, transport, buildings and agriculture could cost-effectively make the transition to a low-carbon economy and which technologies would need to be deployed over time.

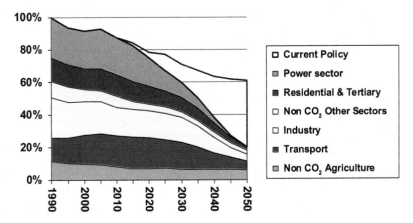

FIGURE 3.1 A cost-effective transition scenario to a low-carbon EU economy in 2050 (greenhouse gas emissions by sector over time as % of 1990 levels)[6]

Source: PRIMES-Gain modelling results, the European Commission's 2050 Low-Carbon Roadmap COM(2011) 112 final

Despite significant variations in the assumptions on technologies and fossil fuel prices, the results turned out to be robust in terms of the overall speed and magnitude of emission reductions. However, more significant variations did show up at sectoral level, depending on the assumptions, for instance in terms of public acceptance of nuclear energy, Carbon Capture and Storage (CCS), projected fuel prices and technological costs over time.

As regards fossil fuel prices, the global analysis showed that these are not just an independent parameter, irrespective of the speed of global climate action. Specifically, it projected a substantial reduction of fossil fuel prices (notably coal and oil) compared to "business-as-usual" projections in the case of global climate action as a result of reduced energy demand and a shift towards low carbon fuels. Global action in a world with relatively lower fossil fuel prices implies that the additional policy efforts such as the required level of (implicit or explicit) carbon prices would have to be stronger. In the case of fragmented action to address climate change, with limited efforts being made by other countries, fossil fuel prices were projected to increase over time, hence making EU action relatively cheaper with larger benefits in terms of fossil fuel imports and energy security. This also demonstrated that for regions such as the EU, which import large amounts of fossil fuels, it does make sense, from a purely economic perspective, to take unilateral action.

The cost-effectiveness analysis as summarised in Table 3.2 shows that the pace of the transition was the fastest in the power generation sector. A much higher penetration of renewable energy, most notably in the power sector, was recognised as an essential building block for decarbonisation. At the same time, the fact that intermittent renewables were projected to massively increase showed the need to consider the implications on market functioning of the power sector, namely the need for more flexible generation and on the network infrastructure required for this transition.

It also showed that the "residential and services" sector would keep up with the general trend, thanks to sustained efforts to build new low-energy housing, deep renovation of existing buildings and increased efficiency of heating and cooling systems, including heat pumps. Industry would do less than average until 2030, mostly through efficiency improvements. However, after 2030, innovative technologies, such as hydrogen or the large-scale deployment of CCS would be needed.

The biggest challenge turned out to be transport, where an increase in emissions of 20% by 2030 could not be excluded, while all other sectors were on a downward path. The analysis suggested that the electrification of the sector was bound to be a crucial element of its decarbonisation. If that were not to happen, a much greater use of biofuels would be needed to achieve the same level of reductions. However, such significant increases in biofuel use, assuming that land-based biofuels would be used, could lead to increased emissions from land use, greater pressures on biodiversity and water management. Today, with the advent of much lower costs for solar, wind and

TABLE 3.2 Sectoral development in the EU for 2050

GHG reductions compared to 1990	2005	2030	2050
Total	−7%	−40 to −44%	−79 to −82%
Sectors			
Power (CO_2)	−7%	−54 to −68%	−93 to −99%
Industry (CO_2)	−20%	−34 to −40%	−83 to −87%
Transport (incl. aviation CO_2, excl. maritime CO_2 and aviation non-CO_2)	+30%	+20 to −9%	−54 to −67%
Residential and services (CO_2)	−12%	−37 to −53%	−88 to −91%
Agriculture (non-CO_2)	−20%	−36 to −37%	−42 to −49%
Other emissions (non-CO_2)	−30%	−71.5 to −72.5%	−70 to −78%

Source: PRIMES-GAINS results, Impact Assessment accompanying the European Commission's 2050 Low-Carbon Roadmap; SEC(2011) 289 final of 08.03.2011

especially batteries, the electrification of the transport sector is happening at a much faster pace than originally envisaged eight years ago and is part of a global trend.

Finally, agricultural emissions are the ones that are reduced the least. This is in large part due to the fact that most of these emissions are intrinsically linked to meat consumption; hence, further reductions would require behavioural changes of diet that were not assumed in the modelling.

3.2.4 Investments

The fourth important finding of the Roadmap related to investment needs. In all decarbonisation scenarios, a massive shift from fuel expenses (operating costs) to investment expenditure (capital expenditure) is observed. This is very important from an economy-wide perspective as investments are to a large extent expenditures in the domestic European economy, requiring increased output and added value from a wide range of manufacturing industries such as automotive, power generation, industrial and grid equipment, energy efficient building materials, etc. Fuel expenditure, on the other hand, is largely flowing to third countries in view of the EU's strong reliance on fossil fuel imports.

To make this low-carbon transition happen, it was estimated that the EU would need to invest an additional €270 billion or 1.5% of its GDP, annually, on average, over the period from 2010–2050 beyond the investment that would be needed anyway. These are largely investments in capital goods, such as low-carbon generation technologies such as solar, onshore and offshore wind, nuclear, CCS, extended grid connections, including smart grids, new automotive and other transport technologies, low-energy houses, more efficient appliances and so on.

The size and composition of low-carbon capital expenditure over the coming decades raised the important question how the increased need for finance could be overcome, in particular for end users of transport and buildings. Innovative finance and fiscal instruments would be required, such as preferential loans, grants and tax rebates, in order to unlock private investment in low carbon technologies. Additionally, a larger share of regional funding within the EU budget would need to go to policy instruments that leverage private sector resources.

These observations led the European Commission to start mainstreaming climate change much more prominently in the European budget, in programmes such as Horizon 2020, the European Structural and Investments Funds (ESIF) and the Connecting Europe Facility. This helped achieve the political decision to "mainstream" climate action in the EU budget from 2014–2020 by setting a target that 20% of all spending should be climate-related.

3.2.5 Innovation

Finally, the Roadmap also made a solid case for a much stronger push for innovation. The development and production of the required low-carbon products provides a major opportunity for Europe's manufacturing industry, provided it succeeds in maintaining and enhancing its technological edge. Europe is no longer blessed with cheap domestic natural resources and has high labour costs, so more innovation is clearly one of the major industrial policy goals needed if the EU is to generate more economic growth and new jobs.

In addition, the EU would become less dependent on expensive imports of oil and gas and less vulnerable to increases in oil prices. On average, and subject to the uncertainties of future oil prices, the EU could save between €175–320 billion annually in fuel costs over the next 40 years. In short, this could free up a tremendous amount of resources for investments in new innovative low-carbon technologies.

Furthermore, greater use of clean technologies and electric transportation such as cars, buses and two-wheelers is expected to substantially reduce air pollution in European cities. Considerably less money would need to be spent on equipment to control air pollution and significant benefits would accrue from reduced mortality, for example. By 2050, the EU could save up to €88 billion a year from air quality benefits.

> Conclusion: The Commission's low-carbon 2050 Roadmap of 2011 showed that a 40% domestic emission reduction by 2030 would be technologically possible and economically feasible. It would require strong carbon pricing, considerable investments and low-carbon innovation in the energy, industry and transport sectors, but it would generate important savings in terms of reduced imports of fossil fuels.

3.3 Towards a Commission proposal for a 2030 climate and energy framework

Despite the analytical quality of the Low-Carbon 2050 Roadmap, the Environment Council failed to agree on Council Conclusions in 2011. Such conclusions are normally agreed by consensus, but Poland was unable to agree to them, essentially because of the fast decarbonisation envisaged in the power sector.

Nevertheless, the 2050 Roadmap was successful in providing guidance on the way forward and in steering the political debate towards a 2030 perspective. Most importantly, the milestone of a 40% domestic reduction in greenhouse gas emissions in 2030 compared to 1990 seemed to bring most Member States together around a potential EU target for the international negotiations leading up to the Paris Agreement. Furthermore, in March 2012, the European Parliament adopted a Resolution[7] endorsing the Commission's Low-Carbon 2050 Roadmap:

> together with its trajectory, the specific milestones for domestic emission reductions of 40 %, 60 % and 80 % for 2030, 2040 and 2050 respectively, and the ranges for sector-specific milestones, as the basis for proposing legislative and other initiatives on economic and climate policy.

The lack of consensus amongst EU Member States on the essential elements of the 2050 Roadmap created a serious political problem. Therefore, the European Commission launched a new process to continue the discussion in two steps, first through a broad consultation exercise and only thereafter through the formulation of an encompassing strategy for 2030.

3.3.1 Green Paper consulting on the possible 2030 framework

In 2013, a consultative "Green Paper" was launched on a potential 2030 climate and energy policy framework,[8] building largely on the main conclusions of the Roadmap and adding some new emerging elements. There was, for example, an increasing recognition that investment in network infrastructure and a flexible electricity market were necessary elements of the low-carbon energy transition. In addition, a number of important changes in the overall global and economic context had taken place that needed to be taken into account. First, there was the fallout of the financial crisis followed by a crisis in the banking sector, leading eventually to budgetary problems in several Member States and to a worsening investment climate. There were also some new emerging trends in global energy markets such as the rapid deployment of renewable energy across the world, increasing competition in low-carbon technologies and the shale gas revolution in the US resulting in increasingly divergent gas prices between the EU and the US.

The Green Paper did not draw conclusions but set out a number of strategic questions, regarding targets, policy instruments, distributional issues, competitiveness, innovation and security of energy supply. While there was wide support for the messages contained earlier in the Low-Carbon 2050 Roadmap, some new elements were also raised such as the advantages and drawbacks of additional targets for renewables and energy efficiency. In light of the substantial response to the consultation, the Commission concluded that there was, "an almost universal support for the development of a common European framework for climate and energy policies."[9] One of the most telling reactions was that of Poland, which suggested that the decision to adopt an objective for 2030 should be taken no earlier than in 2015. Such a message implied that Poland was ready to discuss a climate and energy framework for 2030, but the timing of commitments made should be carefully chosen.

In 2014, a year after the Green Paper, the Commission formulated its strategy in a Communication entitled, "A policy framework for climate and energy policies in the period from 2020 to 2030."[10] Of major importance was a full-fledged economic analysis contained in the accompanying Impact Assessment.[11] It built further on the 2050 Roadmap analysis but added much more specific elements to it.

3.3.2 The Impact Assessment

The Commission analysed different greenhouse gas targets, notably reductions of 35%, 37%, 40% and 45%, in relation to the base year of 1990 and compared these to a reference scenario of there being no new target post-2020. In addition to the 40% scenario, three variants were analysed. First, no extra targets for renewable energy and energy efficiency; second, adding a 30% energy efficiency target; and third, adding a 30% energy efficiency target and a 30% renewable energy target. The 45% greenhouse gas target scenario also included a 35% renewable energy target.

The question whether to have multiple targets for renewable energy and energy efficiency as well as for greenhouse gases received a great deal of attention. If the predominant policy objective were to reduce greenhouse gases, then part of that would be done also through more use of renewable energy or through improvements in energy efficiency, if these avenues proved to be cost-effective. So the political question was whether other energy-related policy objectives would be reached with the definition of one single target for greenhouse gases. If not, what additional costs and benefits would be incurred by setting additional targets?

The results of this analysis are summarised in Table 3.3. A target level of 40% for greenhouse gases would be consistent with an increase of renewables of at

TABLE 3.3 Different levels of greenhouse gases, renewable energy and energy efficiency and their impact on the energy system

	Ref.	*GHG35/ EE*	*GHG40*	*GHG40/ EE*	*GHG40/ EE/RES30*	*GHG45/ EE/RES35*
Main features of the scenarios						
GHG reductions from 1990	−32.4%	−35.4%	−40.6%	−40.3%	−40.7%	−45.1%
Renewables share[1]	24.4%	25.5%	26.5%	26.4%	30.3%	35.4%
Energy savings[2]	−21.0%	−24.4%	−25.1%	−29.3%	−30.1%	−33.7%
Energy system impacts indicators						
Net Energy Imports (2010 = 100)	96	90	89	83	81	78
Energy Intensity[3] (2010 = 100)	67	64	64	60	60	57
Renewables share[4] in electricity, heating and cooling	31.0%	32.6%	34.2%	34.1%	39.7%	47.3%

Source: PRIMES-GAINS modelling results from the European Commission's Staff Working Document: *Impact Assessment accompanying the Communication from the Commission: A policy framework for climate and energy in the period from 2020 up to 2030* (reference: SWD/2014/015 final of 22.1.2014, https://eur-lex.europa.eu/legal-content/EN/TXT/PDF/?uri=CELEX:52014SC0015&from=EN[5]

1 Share of renewable energy in gross final energy consumption according to the 2009 Renewable Energy Sources (RES) Directive.
2 Energy Savings evaluated against the 2007 Baseline projections for 2030.
3 Primary energy to GDP.
4 Contribution of Renewable Energy Sources (RES) in gross final energy consumption of electricity and heating and cooling, based on the individual calculations according to 2009 Renewable Energy Sources (RES) Directive.
5 https://eur-lex.europa.eu/legal-content/EN/ALL/?uri=CELEX:52014SC0015.

least 26.5% and of energy efficiency of at least 25.1%. Similarly, the actually adopted renewable energy and energy efficiency targets for 2030 of 32% and 32.5% respectively imply that, if fully implemented, the greenhouse gas reduction target of 40% would be exceeded and reach a reduction of at least 45% by 2030.

The next issue was to assess the overall economic impacts of such significant change. Table 3.4 shows in different ways that more investments in the energy system would be required and that this figure could sharply increase with the setting of energy targets. However, the reward would be that energy purchases would fall and, in particular, the net imports of fossil fuels. Importantly, no significant impact was expected on average electricity prices unless the greenhouse gas target exceeded 45%. However, a major impact of all this would be reflected in the carbon price. Defining a greenhouse gas target only would result in a carbon price of €40 in 2030, whereas this would fall significantly if binding targets were also set for renewable energy and energy efficiency.

TABLE 3.4 Economic impacts of different scenarios for greenhouse gases, renewable energy and energy efficiency

	Ref.	GHG35/ EE ®	GHG40	GHG40/ EE	GHG40/ EE/RES30	GHG45/ EE/RES35
Economic and social impacts						
Total System Costs, avg annual 2011–30 (bn €)	2,067	2,064	2,069	2,089	2,089	2,102
Total System Cost as % of GDP increase compared to Reference in 2030 in % points	n.a.	+0.03	+0.15	+0.54	+0.54	+0.84
Investment Expenditure[1] in reference and changes compared to reference (avg 2011–30, bn €)	816	+17	+38	+59	+63	+93
Energy Purchases in reference and changes compared to reference (avg 2011–30, bn €)	1,454	−26	−18	−34	−31	−23
Fossil Fuel Net Imports in reference and changes compared to ref. (avg 2011–30, bn €)	461	−10	−9	−20	−22	−27
Average price of electricity[2] (€/MWh) in 2030	176	174	179	174	178	196
ETS price (€/t of CO_2) in 2030	35	27	40	22	11	14

Source: PRIMES-GAINS results, Impact Assessment accompanying the European Commission's 2030 Climate and Energy strategy; SWD/2014/015 final

1 Investment expenditure includes total purchases of transport equipment for households and businesses (including road and non-road transport) but not transport infrastructure costs.
2 Average price of electricity in final demand sectors (€/MWh) constant 2010 Euros. For the reference scenario, the corresponding value was 134 €/MWh in 2010.

3.3.3 The 2014 Commission Communication on the 2030 framework

On the basis of this analysis, the Commission reconfirmed that a greenhouse gas target of at least 40% compared to 1990 was both realistic and economically achievable domestically. It also proposed a "binding" renewable energy target of at least 27%, the level that was projected to be fully consistent with the achievement of the 40% greenhouse gas target. In this way, it aimed to strike a balance between cost-efficiency considerations and the need to continue to provide more policy certainty for investors, as well as give a continued push to the deployment of renewable technologies. Importantly, while this level of "at least 27%" was considered "unambitious" by some, the underlying analysis demonstrated that it would require a share of 45% to 50% renewable electricity in 2030, compared to 30% today.

The Commission refrained from proposing national renewables targets that would be binding on the Member States individually. At this stage, it also refrained from proposing a sub-target for transport[12] and indicated that first generation land-based biofuels made from food and feed should not receive support post-2020. On support instruments for renewables, the need for a more market-driven approach was highlighted. Support schemes needed to be rationalised to become more cost-effective and coherent with the internal market and provide greater legal certainty for investors.[13] These principles were the inspiration for a subsequent revision of the guidelines on state aid for environmental protection and energy 2014–2020,[14] which has led to tendering systems becoming the standard approach for renewable electricity support.

The Commission at the time refrained from proposing a target for energy efficiency in view of a pending review of the Energy Efficiency Directive. However, later in 2014 it proposed a 30% improvement in energy efficiency by 2030, which has since been increased to 32.5%. This recognised the higher costs that would result but sought important benefits in terms of security of supply, notably gas imports in light of the annexation of Crimea and the continued troubles between Ukraine and Russia.

The overall target of 40% greenhouse gas reduction still needed to be distributed in terms of the policy effort between the sectors covered by the harmonised EU Emissions Trading System (EU ETS) and those sectors for which Member States were responsible through the Effort Sharing Regulation[15] (see Chapter 5). The results in Table 3.5 show that, based on cost-effectiveness considerations, the EU ETS sectors would have to deliver a reduction of 43% in greenhouse gas emissions (implying a change in the cap

TABLE 3.5 Cost-effective distribution of 2030 effort between EU ETS and non-ETS sectors

	Ref.	GHG35/ EE ®	GHG40	GHG40/ EE	GHG40/ EE/RES30	GHG45/ EE/RES35
Environmental impact indicators						
GHG emissions reduction in ETS sectors vs 2005	−36%	−37%	−43%	−38%	−41%	−49%
GHG emissions reduction in non-ETS sectors vs 2005	−20%	−26%	−30%	−35%	−33%	−34%
Reduced pollution control & health damage costs (€bn/year)[1]		3.8–7.6	7.2–13.5	17.4–34.8	16.7–33.2	21.9–41.5

Source: PRIMES-GAINS results, Impact Assessment accompanying the European Commission's 2030 Climate and Energy strategy; SWD/2014/015 final

1 Reduction of health damage costs due to reduced air pollution compared to the reference (€bn/yr). Valuation uses value of life year lost used for the Thematic Strategy on Air Pollution, ranging from €57,000 to €133,000 per life year lost.

by adjusting the Linear Reduction Factor from 1.74% annual reduction to 2.2% from 2021) and the non-ETS sectors a reduction of 30%, both compared to 2005. The environmental analysis also reconfirmed the important secondary benefits that could be attained through an improvement of air quality.

As regards the EU ETS, the Communication was accompanied by a specific legislative proposal to introduce a so-called "Market Stability Reserve" to remove the structural surplus that had been accumulating in the EU ETS. This is explained more fully in Chapter 4. In this way, the Commission maintained that the EU ETS could be restored as a central, EU-wide and cost-effective driver for low-carbon investment for larger emitters. Notable is that the definition of the "surplus" would be done regardless of its cause, which could be the result of slowing or declining economic activity, through more production of renewable energy or higher improvements in energy efficiency. With such a Market Stability Reserve, the risk for potential negative spillovers from national sectoral policies on the functioning of the EU ETS could be largely neutralised.

As regards the other non-ETS sectors, the Commission proposed to continue the 2020 approach to share the effort between Member States based on GDP *per capita*, with more wealthy countries taking on more ambitious targets. This was based on analytical evidence in the Impact Assessment that less wealthy, less efficient and more carbon intensive countries were facing relatively higher investment requirements. At the same time, flexibility between Member States in these sectors would help to ensure that efforts were made where they would be most cost-effective.

By specifying not only the headline climate target but also the accompanying architecture with EU-level targets for renewable energy and energy efficiency, the Commission wished to facilitate the reaching of consensus on the key elements of the 2030 climate and energy framework before making the specific legislative proposals. It was expected that such agreement reached before the end of 2014 would later on facilitate the future legislative process, with respect to both the EU ETS and the Effort Sharing legislation for 2030.

3.3.4 The European Council Conclusions of October 2014

With the gradual development of EU climate policy it was becoming clear that policy making was not becoming simpler, even if most other countries across the world were giving strong signals that they were ready to come together to act. It was remarkable that the Environment Council – bringing together all Ministers for the Environment of the 28 EU Member States – had, in 2011, failed to agree on the way forward in a 2030 perspective. In contrast, by 2014, sufficient political momentum had been created in anticipation of COP20 in Warsaw and COP21 to be held in Paris in 2015.

The normal institutional way of proceeding would have been for the Commission first to make a set of policy proposals. However, without a strong political mandate endorsed by the highest levels of government, these could have taken a long time to be agreed upon, or, worse, could have been rejected. The first European Council President, H. Van Rompuy chose a much more astute way. He decided to bring the essential elements of the Commission's 2030 energy and climate framework to the Heads of State and Government. On 23–24 October 2014, the European Council unanimously adopted a set of conclusions regarding the 2030 climate and energy framework, including the proposed binding domestic target of at least 40% greenhouse gas reductions compared to 1990.[16] This provided clear political guidance for the further legislative proposals the Commission would have to propose later.

This result in October 2014 was of key importance for the international negotiations because it allowed the EU to submit its "Intended Nationally Determined Contribution," in line with the timeline agreed by the Parties to the UNFCCC, at COP20 in Warsaw for the conclusion of an international agreement (later to become the Paris Agreement). The EU's early submission had its effect on other Parties too, serving as an incentive to make their contributions. Less than a month later, on 12 November 2014, US President Obama and Chinese President Xi jointly announced,

> their respective post-2020 actions on climate change. . . . The United States intends to achieve an economy-wide target of reducing its emissions by 26%-28% below its 2005 level in 2025 and to make best efforts to reduce its emissions by 28%. China intends to achieve the peaking of CO_2 emissions around 2030 and to make best efforts to peak early and intends to increase the share of non-fossil fuels in primary energy consumption to around 20% by 2030.

Apart from the headline 40% reduction target, the European Council had to agree on a range of other issues in order to find a political consensus. It was agreed that the Linear Reduction Factor of the EU ETS would be increased from 1.74% per year to 2.2% per year from 2021. This would equate to a reduction of emissions from the EU ETS sectors of 43% compared to their level in 2005. In exchange for this significant strengthening of the ambition, the existing system of free allocation would be continued in order to avoid potential risks related to a relocation of energy intensive industry outside Europe. This was an important decision as the EU ETS legislation at the time foresaw a phasing out of free allocation as of 2021. For the sectors not covered by the EU ETS, a reduction of emissions of 30% compared to 2005 was agreed upon for the EU as a whole and that differentiated reduction targets amongst Member States would vary within a range between 0% and −40% compared to 2005.

This was all consistent with the Commission's Communication "A policy framework for climate and energy in the period from 2020 to 2030" of January 2014. However, it was clearly necessary to forge additional compromises on key issues in order to reach consensus. These compromises outlined specific conditions for the ensuing climate legislation. In contrast to the 2020 package, these elements were now tackled head-on before the Commission tabled its specific legislative proposals. It must be admitted that this limited the room for manoeuvre for the Commission in presenting its legal proposals, as well as excluded the European Parliament from this early stage of policy design. However, even though the rules of the ordinary legislative procedure applied in full to the specific legislative proposals, this did unquestionably facilitate adoption of the legislation and avoided the need for the European Council to become directly involved in the legislative process later on.

As regards the distribution of effort, it outlined a delicate balance regarding the contribution mostly between less wealthy and wealthier countries, or in other words, how a cost-effective implementation should be reconciled with elements of fairness and solidarity, including the following elements:

1   The creation of a Modernisation Fund for lower-income Member States (below 60% of average GDP *per capita*) from auctioning revenues;
2   The distribution of 10% of auctioning revenues to countries with GDP *per capita* below 90% of the EU average;
3   Effort for the non-EU ETS sectors to be based on GDP *per capita* ranging between 0% (for the least wealthy) and −40% (for the wealthiest) as compared to emission levels in 2005;
4   Targets for the Member States with a GDP *per capita* above the EU average to be adjusted to reflect cost-effectiveness in a fair and balanced manner;
5   A limited "one-off" transfer of EU ETS allowances to the non-ETS sectors providing flexibility for the Member States with targets both above the EU average and above their cost-effective potential.

The European Council further endorsed a target of "at least 27%" for renewable energy "binding at EU level" and an "indicative target" of "at least 27%" for energy efficiency, to be reviewed by 2020, "having in mind an EU level of 30%." For renewable energy and energy efficiency, the European Council did not endorse the idea of setting nationally differentiated binding targets, once again in line with the Commission's initial Communication.[17]

For renewable energy it was indicated in the European Council conclusions of October 2014 that Member States can set more ambitious targets "and support them in line with State aid guidelines," thereby reassuring Member States, such as Germany, that wished to continue strong national policies and support systems. No sub-target was set for transport, unlike for 2020, reflecting at that time the wish of Member States for greater flexibility.

An additional objective of "arriving at a 15% target by 2030" with respect to interconnectivity in electricity networks between Member States was also agreed upon. This reflected the crucial role of electricity connectors to strengthen the EU internal market for electricity, to enable greater penetration of renewable energy and to improve security of supply. The crucial role of infrastructure to the proper functioning of the internal energy market was understood, including the important contribution to be made by co-financing projects with contributions from the EU budget.

Finally, the European Council endorsed the idea of an Energy Union governance system to bring together existing national climate, energy efficiency and renewable plans into one integrated national plan, to enable a systematic monitoring of key indicators and to facilitate regional cooperation.

Conclusion: The Commission formulated a comprehensive strategy for 2030, based on the Low-Carbon 2050 Roadmap. The European Council, bringing together the EU's Heads of State and Government, endorsed this strategy and adopted the "at least 40%" target for 2030. This injected a welcome impetus into the international preparation of the Paris Agreement.

Conclusion

It was a long and complicated journey to obtain a consensual agreement between Member States on a long-term strategy for EU climate policy and even more so to agree on a specific EU-wide reduction target for 2030. The European Council, as mandated by the Lisbon Treaty, turned out to be a necessary and helpful facilitator in reaching an agreement.

This new institutional dynamic enabled the European Union to maintain its international leadership in the UN negotiations on climate change by assuming an ambitious quantitative emissions cap. As such, it set an example for other countries and maintained the momentum behind submissions of Intended Nationally Determined Contributions in the run-up to COP21 in Paris.

The Commission put thorough economic analysis at the heart of its approach to maintaining fairness between Member States. The economic modelling work informed the Heads of State and Government and facilitated the subsequent legislative work of the Council and European Parliament relating to climate change and the energy transition. Nevertheless, it took seven years of continuous effort between the formulation of the Low-Carbon 2050 Roadmap in 2011 and adoption in 2018 of the legislation implementing the EU's climate and energy targets for 2030, in particular the Effort Sharing Regulation, the amendments to the EU ETS, the Energy Efficiency and the Renewable Energy Directives.

Notes

1 European Council, Presidency Conclusions – Brussels 8–9.3.2007, Council of the European Union, 7224/1/07, 2.5.2007.
2 COM(2011)112 final of 08.03.2011, Communication. *A Roadmap for Moving to a Competitive Low Carbon Economy in 2050*. http://eur-lex.europa.eu/resource.html? uri=cellar:5db26ecc-ba4e-4de2-ae08-dba649109d18.0002.03/DOC_1& format=PDF.
3 COM(2011)885 final of 15.12.2011: Communication: "Energy Roadmap 2050". http://eur-lex.europa.eu/legal-content/EN/TXT/PDF/?uri=CELEX:52011D C0885&rid=3.
4 Domestic reductions would exclude international offsets being used to achieve emission reduction objectives.
5 Climate targets for the first commitment period of the Kyoto Protocol and for 2020 included the use of international offset credits.
6 COM(2011)112 final, 8.3.2011.
7 European Parliament resolution of 15 March 2012 on a Roadmap for moving to a competitive low carbon economy in 2050. http://www.europarl.europa.eu/sides/getDoc. do?pubRef=-//EP//TEXT+TA+P7-TA-2012-0086+0+DOC+XML+V0//EN.
8 COM(2013)169 final, Green Paper. "A 2030 Framework for Climate and Energy Policies". http://eur-lex.europa.eu/legal-content/EN/TXT/PDF/?uri=CELE X:52013DC0169&from=EN.
9 Commission's Services Non-paper, Green Paper 2030: Main Outcomes of the Public Consultation. https://ec.europa.eu/energy/sites/ener/files/documents/ 20130702_green_paper_2030_consulation_results_1.pdf.
10 COM(2014)15 final. "A Policy Framework for Climate and Energy in the Period from 2020 to 2030". http://eur-lex.europa.eu/legal-content/EN/TXT/PDF/?u ri=CELEX:52014DC0015&from=EN.
11 SWD (2014)15 final, Commission Staff Working Document Impact Assessment accompanying. "A Policy Framework for Climate and Energy in the Period from 2020 to 2030".
12 In the end, the new renewable energy Directive (EU) 2018/2001, adopted in 2018, sets an overall renewable energy target of "at least 32%" in 2030 and 14% by 2030 for renewable energy in the transport sector (see Official Journal L328, 21.12.2018, pp. 82–209).
13 The practice of retroactive changes to renewable energy support measures in some countries had undermined investor confidence and led to a stand-still in investments in those countries.
14 Official Journal C 200/1 of 28.6.2014, pp. 1–55.
15 "Regulation (EU) 2018/842 of the European Parliament and of the Council of 30 May 2018 on binding annual greenhouse gas emission reductions by Member States from 2021 to 2030 contributing to climate action to meet commitments under the Paris Agreement and amending Regulation (EU) No 525/2013", OJ L 156, 19.6.2018, pp. 26–42. See: https://eur-lex.europa.eu/legal-content/EN/ TXT/PDF/?uri=CELEX:32018R0842&from=EN.
16 European Council Conclusions EUCO 169/14 of 24.10.2014.
17 However, unlike the climate elements, these renewable energy and energy efficiency targets were revised later in the legislative negotiating process between the Council and the European Parliament on the specific renewable energy and energy efficiency legislation (see Official Journal L328 of 21.12.2018, pp. 82–230 respectively).

4

THE EU EMISSIONS TRADING SYSTEM

Damien Meadows, Peter Vis and Peter Zapfel

Introduction

The development of the EU Emissions Trading System (EU ETS) up to 2014 is comprehensively elaborated in *EU Climate Policy Explained*,[1] while *EU Energy Law: The EU Emissions Trading Scheme*[2] dealt with the origins of the EU ETS and its early years of operation. This chapter reviews what the EU ETS has achieved so far in terms of emission reductions. It further focuses on recent developments, relating in particular to the legislation covering the period from 2021–2030. This legislation implements a significant share of the EU's target to reduce its greenhouse gas emissions by "at least 40%" by 2030.

4.1 How does the EU Emissions Trading System work?

The EU ETS is a "cap-and-trade" system that guarantees an environmental outcome by setting a cap on the total amount of carbon emissions. The quantity of allowances issued serves as the quantitative cap on emissions, and these allowances are then either auctioned or allocated for free to companies. Companies have an obligation to regularly hand over to governments sufficient allowances to cover their actual emissions. Companies may trade these allowances. Progressively, the total number of allowances in the system is reduced at a steady and predictable rate. This secures an improved environmental outcome over time. This is a "cap" on emissions, and certainly not a "cap on growth," as is sometimes wrongly claimed by critics. Historical data shows that economic activity covered by the system has grown collectively even while emissions have been coming down.

The advantage of a market-based system is that it incentivises reductions in emissions across all entities covered by the system in a cost-effective manner. Companies have an economic interest to cut emissions and sell allowances when the market price for allowances is higher than the cost of reducing its own emissions. Conversely, companies with reduction costs above the market price are able to buy allowances. This means that, across the system, there is an incentive for reductions to take place where the costs of abatement are lower, while the environmental outcome remains guaranteed by the overall emissions cap. By covering a variety of sectors, the EU ETS allows continued growth in emissions from individual sectors by buying allowances from other sectors where emission reductions are cheaper.

By putting a price on carbon, a market failure is corrected, and companies and economic actors are incentivised to take account of this in their operational decision-making and long-term investment planning. Carbon prices strengthen the business case for making investments in low-carbon technology: the rate of return is improved, and the payback period reduced compared to more carbon-intensive alternative investments. Putting a price on carbon is therefore an important signal for the economy. Moreover, it must be made very clear to all, particularly higher emitting sectors, that carbon allowances will continue to reduce significantly over time.

Emissions trading systems, as well as other market-based measures like carbon taxes, have the potential to generate money that can be used for climate change mitigation and adaptation. Polluters then not only have to pay to pollute, but the revenues generated can then be redeployed to further stimulate innovation and deployment, or to address societal effects of climate constraint such as retraining employees in carbon-intensive industries like coal mining.

The EU ETS works with the economic cycle: for example, a recession leads to lower emissions, affecting the supply/demand balance in the carbon market and causing a lower carbon price, while an economic recovery could have the reverse effect. A fluctuating carbon price is a normal feature that does not undermine the overall predictability of the EU ETS. Companies can save emission allowances until they need them or sell allowances they do not need. This flexibility gives an incentive to reduce emissions earlier in time and for individual companies to overachieve. Other companies have the flexibility to buy additional allowances if they find making investments themselves is too expensive.

A well-functioning market requires the trust and confidence that actors will comply with the rules. The EU ETS therefore developed a solid system of monitoring, reporting, verification (MRV) and compliance. This is essential for a market-based measure to work. Since 2008, in the EU ETS, in the case of failure to comply, there has been an inflation-linked penalty of €100 per tonne of excess emissions plus the obligation to make up the shortfall.[3]

> Conclusion: The EU ETS puts a price on carbon. It provides incentives to companies to reduce emissions and ensures that the cap set on their collective emissions is met in a cost-effective manner.

4.2 Emission reductions of 26% under the EU ETS from 2005–2017

Today, the EU ETS covers 40% of the EU's CO_2 emissions and is therefore critical to the delivery of the overall climate targets. It regulates, in a harmonised way, emissions from some 11,000 installations, mainly from electricity and heat generation, manufacturing industry and around 500 airlines for their intra-European flights.

Between 2005 and 2017, the emissions covered by the EU ETS fell by 26%, which is more than the average for the EU as a whole (Figure 4.1). In other words, the EU ETS sectors are doing more than other sectors are doing. Emissions covered by the EU ETS have reduced from over 2 billion tonnes per year in 2008 to less than 1.7 billion tonnes per year in 2017.[4] The EU ETS has continuously ensured reductions in emissions while maintaining a very high level of compliance.

Due to the lack of monitoring at installation level and independent verification, no comparable figures exist for the years prior to the introduction of the EU ETS in 2005. However, several studies[5] point to the fact that the carbon price signal has resulted in real emission reductions since the very beginning of the EU ETS. The largest drop in emissions happened, however, between 2008 and 2009, which was to a considerable extent influenced by the onset of the economic crisis in late 2008.

Since 2013, taking into account extensions of scope, there has been an average yearly reduction in the EU ETS sectors of around 2.6%. The power sector delivered the highest reduction by an average reduction of more than

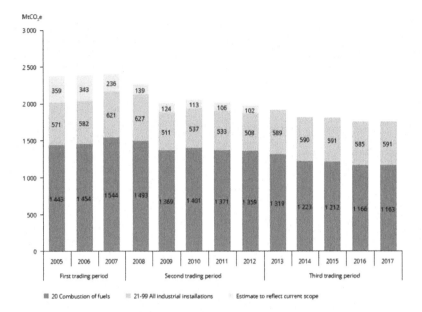

FIGURE 4.1 Verified emissions under the EU ETS, 2005–2017

Source: European Environment Agency (EEA), "Trends and projections in the EU ETS in 2018" (page 21). Note that the estimate to reflect current scope takes in account additional emissions (not split by activity) for the period from 2005–2012 to provide a consistent time series for the coverage of emissions in the third trading period.

4% per year since 2013, reflecting fuel switching away from coal and an increased use of renewable energy sources. Emissions from industrial sectors reduced as well, but to a much more limited extent of around 0.5% per year. Unfortunately, no robust empirical studies are yet available indicating the impact of specific policy instruments in driving these emission reductions.

Contrary to the significant reduction in stationary sources, the emissions from intra EU aviation have been growing by some 5% per year,[6] up to some 67 million tonnes per year in 2018.[7] The annual free allocation is around 30.5 million allowances, based on airlines' efficiency in transporting passengers and cargo, while around five million allowances were auctioned. Airlines are therefore buying more than 30 million allowances from other sectors every year to offset the growth in their emissions, alongside a small proportion of offset credits from the Kyoto Protocol's Clean Development Mechanism (CDM) that continue to be allowed up to the year 2020.

Conclusion: The EU ETS covers 40% of EU's greenhouse gas emissions, creating a clear and predictable reduction of the emissions cap over time. Since 2005, emissions covered by the EU ETS decreased by 26%.

4.3 Addressing the low carbon price since 2013

The EU ETS operates in phases, as for each phase significant modifications were made to the legislation. Phase-1 from 2005–2007 was considered to be a pilot phase, as all institutional infrastructure needed to be created, even before the Kyoto Protocol started. Phase-2 covered the years 2008–2012 and corresponded with the first commitment period under the Kyoto Protocol. Phase-3 covered 2013–2020, which is currently running and was intended to become Kyoto's second commitment period. Finally, Phase-4 covers the period from 2021–2030 that corresponds to the European Union's first commitment under the Paris Agreement. In Figure 4.2, the prices of allowances issued in each phase are shown in different colours.

In the first phase, the value of EU allowances for that phase dropped steeply in 2006, when the first verified emissions figures were reported. This reflects the fact that the EU ETS began in 2005 without a detailed database of actual emissions per installation, and it became clear that the

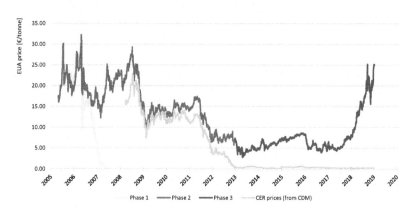

FIGURE 4.2 Price trends for allowances under the EU ETS and EU ETS-eligible international credits (CER) under the Kyoto Protocol (€/tonne)

Source: ICE (for EUA prices), Refinitiv (for CER prices, based on EEX and ICE price data reports 2013–2018)

Member States were issuing a number of allowances in excess of expected total emissions. As these allowances could not be banked into subsequent phases, this excess resulted in a price of nearly zero in 2007. At the same time, the market price in 2007 for allowances valid for 2008–2012 was much higher in view of expectations that the system would be more constrained in the future.

A second major price drop came at the end of 2008, as the scale of the global economic and financial crisis became clear. It is important to note that sectors covered by the EU ETS in aggregate were subject to much stronger swings in output than the economy as a whole. Individual sectors covered by the EU ETS had output drops between 2008 and 2009 of over 30%, and the supply of allowances started to exceed demand.

By 2012, the market had built up a supply overhang of almost two billion allowances. This over-supply was significantly aggravated by a sizeable inflow of international credits in 2011 and 2012 especially (see Figure 4.3). Indeed, the EU ETS allowed for the use of over 1.5 billion tonnes of international credits created under the Kyoto Protocol. At the same time, the EU was deliberately promoting energy efficiency improvement and the increased use of renewable energy, but no correction of the cap for these policies was foreseen. All these factors, including the economic recession, the international credits and the effects of energy policies, led to a growing market imbalance weighing heavily on the price that fell to single digits.

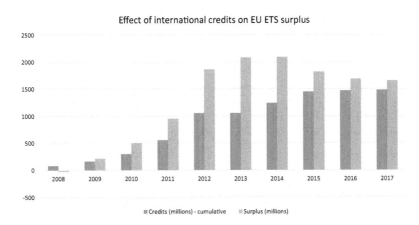

FIGURE 4.3 Surplus of EU ETS allowances with cumulative number of CERs used in EU ETS

Source: European Commission

While emissions under the EU ETS kept falling, the carbon price of around €5 per tonne prompted a policy debate on how to restore market confidence and rejuvenate the European carbon market as a driver for low-carbon investments. As the political environment did not allow for a tightening of the cap, a short-term legislative response was developed. It was decided to reduce the quantity of allowances for auctioning in 2014–2016 by 900 million allowances[8] but to bring them back to the market at the end of Phase-3 (a mechanism referred to as "back-loading"). This made the surplus shrink in the short-term by more than 40%, but this was only a temporary solution.

In the meantime, a longer-term response was developed through the creation of a "Market Stability Reserve," adopted in 2015.[9] This Market Stability Reserve provides for an automatic lowering of the auction volume when the cumulative surplus of allowances in the market exceeds 833 million allowances, with the consequence that a pre-determined number of allowances would be put into the Reserve. Allowances will be released automatically from the Reserve at a pre-determined rate once the cumulative surplus of allowances in the market falls below 400 million allowances. The Market Stability Reserve is intended to act like a dry sponge that absorbs surplus allowances in times of over-supply and releases allowances in times of under-supply as if the sponge is being squeezed. The purpose is to maintain levels of liquidity needed for the carbon market to function properly, estimated to be between 400 and 833 million allowances, to cover both "spot" and "futures" transactions.

The rules of functioning of the Market Stability Reserve are automatic and pre-determined. The aim is to avoid discretionary interventions and to be as predictable as possible, hence creating market stability. The thresholds and the rate managing the inflows and outflows to and from the Reserve are set in the Directive. The principle is that as the supply and demand balance are automatically adjusted, the market is left to determine the price – as is generally intended with market-based instruments. The Market Stability Reserve is operating from 2019. The 900 million back-loaded allowances taken out of the market between 2014 and 2016 will be put directly into the reserve, together with a proportion of the EU ETS surplus allowances in accordance with the operating rules of the Reserve.

It is important to note that the definition of the surplus is defined irrespective of how such a surplus originates, whether due to a business cycle downturn or because of energy policy measures such as on coal plant shutdowns, renewable energy expansion or energy efficiency measures. In this way, the Market Stability Reserve puts an end to questions related to the compatibility of the EU ETS and other policies, in particular in the energy sector. The synergy between the EU ETS and other policy measures has been improved considerably through the creation of the Market Stability Reserve; this is of critical importance for the coming decade where the role of renewable energy is expected to grow considerably.

The Market Stability Reserve provides the EU ETS with an instrument that introduces a volume-based adjustment mechanism. The European legislator opted not to set minimum and maximum prices for allowances, which would constitute a "price collar." This could have resulted in mixing price and volume determination simultaneously and could have led to unpredictable results.

The experience of the carbon market in the exceptional turbulent economic times did require more corrections than originally foreseen, but these were necessary to safeguard the proper functioning of the system. This was just another expression of what happened in the financial markets more generally following the deepest financial and economic crisis since the Second World War. These corrections were made through the normal decision-making process and therefore required time. The result is that the EU ETS is now more predictable and shock-resilient than it has ever been, and the likelihood of any need for future corrections is now considerably reduced.

> Conclusion: Between 2010 and 2017 the European carbon price was low, largely as a result of the economic recession, a large influx of international credits and developments in energy markets. The Market Stability Reserve makes the EU ETS more resilient to future major changes in the economy or in energy policy.

4.4 A fundamental review for the period from 2021–2030

Extensive economic analysis performed in the context of the EU's overall 40% reduction commitment by 2030 has shown that an emissions reduction of 43% below 2005 represents a cost-effective contribution from the EU ETS sectors. This corresponds to an increase of the Linear Reduction Factor of the EU ETS from 1.74% to 2.2% per year from 2021 onwards. This is an essential component of the EU's long-term strategy, and this annual linear reduction continues to apply by default after 2030.

During the Phase-4 negotiations, the question arose of whether the increase in the linear reduction to 2.2% and the creation of the Market Stability Reserve would be sufficient to address the oversupply on the carbon market. In the end, it was agreed to double the rate at which allowances are placed into the Market Stability Reserve from 12% to 24% between 2019 and 2023. In addition, unless decided otherwise in the first review of the Market Stability Reserve in 2021, a time limit was introduced on the validity of any allowances in the Market Stability Reserve in excess of the level of allowances auctioned during the previous year. From 2023, any allowances in the Market Stability Reserve above this level will no longer be valid and this limitation will continue to apply on an annual basis if the surplus exceeds the level of allowances auctioned during the preceding year. As a result, from 2023 the size of the Market Stability Reserve will be limited to the amount of allowances auctioned in the previous year.

There is now also a specific provision that recognises Member States may cancel allowances that would otherwise have been auctioned, in the event that a Member State phases-out the use of coal or lignite fuel in the power sector. Such phase-out policies in electricity generation have the potential to cause substantial surpluses of allowances in the market, thereby weakening the incentive to reduce emissions elsewhere.

After the formal agreement of the EU ETS reforms for Phase-4, the carbon price returned to its pre-2010 level (see Figure 4.2). This happened quicker than expected.

All these changes combined have fundamentally influenced the outlook on the supply and demand of allowances until 2030 (see Figure 4.4). The surplus in the market, built up in the years following the financial crisis of

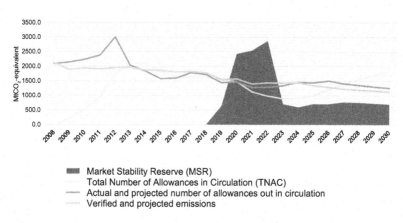

FIGURE 4.4 Outlook on the supply and demand of allowances until 2030

Source: ICIS, February 2019

2008, will first be absorbed in the Market Stability Reserve and after 2023 be neutralised. Consequently, the normal scarcity for a well-functioning market mechanism is gradually reappearing.

> Conclusion: The adoption of legislation on Phase-4 of EU ETS led to the return of the price to its pre-recession levels. Along with a reinforced role of the Market Stability Reserve, the total amount of allowances will further reduce by 2.2% annually from 2021, an increase of the previous applicable rate of 1.74% in the period from 2013–20.

4.5 Carbon leakage and free allocation: having industry on board

Economists usually advocate that all allowances should be auctioned in accordance with the polluter-pays principle, a principle explicitly mentioned in the EU Treaty. However, this could lead to negative impacts on the competitiveness of European companies if done while other major economies are not putting a similar price on the external costs of emissions or taking other comparably stringent actions to reduce greenhouse gas emissions.

Emission trading is very transparent in terms of its price signal and, while recognising that there are many factors involved in investment and operational decisions, an important political issue is not to risk losing industrial production from Europe to other countries. Relocation from Europe could theoretically lead to an increase in global emissions if the technology used in these economic activities would not be as carbon-efficient as in Europe (referred to as "carbon leakage").

In this context, free allocation allows for the avoidance of adverse impacts on competitiveness. Under the current Phase-3, over 40% of allowances under the EU ETS are therefore given out for free. Since 2013, the rules for free allocations have been harmonised across the EU ETS to make sure that companies are treated the same way irrespective of the Member State they are established in.

The Heads of State and Government decided in 2014 to continue the current approach, contrary to the foreseen complete discontinuation of free allocation from 2027. Consequently, it was decided that in principle 43% of allowances would be handed out for free between 2021 and 2030, but in certain circumstances this amount could increase to 46%. It would be important, however, to make sure that these free allowances are given to those sectors and economic activities where the risk of carbon leakage is real.

4.5.1 Benchmarks

In total, 54 technological benchmarks have been adopted. Given the diversity in the manufacturing sector, it is not possible for every product to have a specific benchmark, but 52 key product benchmarks have been established and cover the major part of industrial emissions. The remainder are covered either by the application of the heat-based energy benchmark, or to a minor extent by a fuel-based energy benchmark. Finally, a very small percentage of allowances are allocated in relation to a process emissions rule based on past emissions levels.[10] There is a separate benchmark for aviation activities, described in Chapter 7. Table 4.1 gives

TABLE 4.1 Examples of initial benchmark values for 2013–2020 (tonne CO_2 per 1000 tonnes of output produced)[1]

Product	Benchmark	Product	Benchmark	Product	Benchmark
Coke	286	Sintered ore	171	Hot metal	1328
Pre-bake anode	324	Aluminium	1514	Grey cement	766
White cement clinker	987	Lime	954	Dolime	1072
Sintered dolime	1449	Floatglass	453	Bottles & jars colourless	382
Bottles & jars coloured	306	Continuous filament glass fibre	406	Facing bricks	139
Pavers	192	Roof tiles	144	Spray dried powder	76
Plaster	48	Dried secondary gypsum	17	Short fibre kraftpulp	120
Long fibre kraftpulp	60	Sulphite pulp	20	Recovered paper pulp	39
Newsprint	298	Uncoated fine paper	318	Coated fine paper	318
Tissue	334	Testliner and fluting	248	Uncoated carbon board	237
Coated carbon board	273	Nitric acid	302	Adipic acid	2790
Vinyl chloride monomere	204	Phenol/ acetone	266	S-PVC	85
E-PVC	238	Soda ash	843		

Source: Annex I of Commission Delegated Regulation (EU) 2019/331 of 19 December 2018, determining transitional Union-wide rules for harmonised free allocation of emission allowances pursuant to Article 10a of Directive 2003/87/EC of the European Parliament and of the Council.

1 For benchmarks defined without consideration of exchangeability of fuel and electricity.

the benchmark values that formed the initial reference point for free allocation for 2013–2020, and which are the starting points for calculating the trajectories for benchmark value improvements for Phase-4. These benchmark values took account of the most efficient techniques, substitutes and alternative production processes. All allocations were decided prior to the period in question, and benchmarks are calculated on the basis of past production quantities rather than inputs to the production process,[11] in view of maximising incentives for emissions reductions and energy efficiency savings. Most product benchmark values were derived from the average performance of the 10% most efficient installations in a sector in the EU in 2007–2008, based on data from all EU Member States.

The initial setting of the benchmark values was a complicated process, partly because of different industrial strategies followed by different companies in the same sector. Therefore, it is not entirely surprising that the benchmarking decision and the free allocation process led to a number of legal challenges, although the judgements upheld the validity of the benchmark values in all cases.[12]

4.5.2 Carbon leakage list

A wide range of industries are included on a list of sectors "deemed to be exposed to carbon leakage," and they receive allocation at the level of 100% of the harmonised benchmarks. Industrial facilities not covered by this status are allocated 80% of the benchmark in 2013, declining annually and in a linear manner to a level of 30% from 2020–2026, and they will then reduce to zero between the years 2027–2030.

The list of industries "deemed to be exposed to a significant risk of carbon leakage" was first adopted in 2009[13] for five years, and a second list was adopted for 2015–2020.[14] A sector was "deemed to be exposed to a significant risk of carbon leakage" if the sum of additional costs related to both the direct emissions and the indirect impacts from the use of electricity would lead to an increase of production costs of 5% or more and the sector's intensity of trade with third countries was above 10%. Sectors were also included if either EU ETS direct and indirect additional costs would lead to an increase of production costs of at least 30%, or the sector's intensity of trade with third countries exceeds 30%. Most of the sectors and sub-sectors were included on the 2009 list because their intensity of trade with third countries exceeded 30%.[15] Other sectors were included on the list based on a qualitative assessment, taking into account the extent to which it is possible for installations to reduce emission levels or electricity consumption, current and projected market characteristics and profit margins.

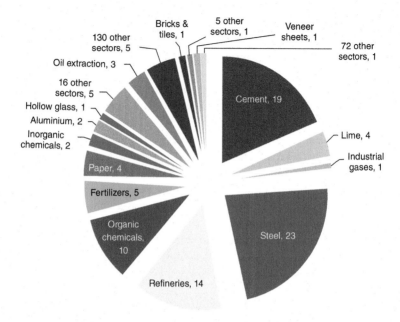

FIGURE 4.5 Share of free allocation (%) based on carbon leakage list, 2013–2020

Source: European Commission, based on Figure 11 in Impact Assessment for EU ETS amendment proposal 2015, reference SWD(2015)135 final of 15.07.2015[16]

Based on the harmonised benchmarks and the carbon leakage list, Member States calculate in advance the number of free allowances for each installation based on the EU-wide rules. So far, the predominant part of the free allocation goes to only a limited number of sectors such as cement, steel and chemicals as shown in Figure 4.5. Over the period from 2021–2030, some 6.3 billion allowances will be given out for free out of a total of 15.5 billion allowances.

4.5.3 Update of the carbon leakage list and benchmark values for Phase 4

For Phase-4, differentiation is maintained between sectors that are exposed to a significant risk of carbon leakage, on the one hand, and other industry sectors on the other hand. Whether a sector is included on the carbon leakage list is determined in general based on a single criterion[17] that reflects both the mathematical outcomes of carbon intensity and trade intensity of the sector.[18]

This limits the number of sectors included on the list to 50 sectors and 12 sub-sectors. This list will be applicable for the whole period from 2021–2030.

For the period from 2021–2030, a combined data collection from thousands of industrial installations will provide the basis for the update of the benchmark values and the determination of the production levels needed for free allocation purposes. As from 2021, the existing benchmark values will be updated to reflect technological progress within a range of 0.2% and 1.6% per year, applicable between 2008 and the middle of each five-year period in Phase-4. This ensures that financial incentives for those sectors experiencing more rapid technological progress and emission reductions are maintained. By way of exception, the benchmark value for hot steel is only to be updated by the lower rate of 0.2% for allocations in 2021–2025. The reductions for each benchmark value will therefore range between 3–24% of the initial benchmark value for allocations during 2008–2023, and between 4–32% of the initial benchmark value for allocations during 2026–2030.

On the manner of adjusting the 54 benchmark values, the revised Directive empowered the Commission to adopt a "delegated act" concerning the union-wide and fully harmonised rules for the allocation of free allowances, which has been done through the Free Allocation Regulation.[19] This regulation determines detailed implementation rules including on definitions, modifications to the monitoring rules regarding the data collected by Member States, the production level data and the determination of historical activity levels.

Allocations will also be adjusted if installations' operations increase or decrease by more than 15%, in order to align free allocations more closely with actual production levels.[20] An implementing act on allocation adjustments is being prepared for adoption in 2019.

These updated benchmarking rules are important, as they introduce a pioneering step to have a rule-based process to update the benchmarks periodically.

4.5.4 The correction factor

Since 2013, a safeguard clause ensures that the amount of allowances given out for free on the basis of benchmarks does not exceed the amount of allowances available for free within the overall cap.[21] In respect of 2013–2020, this maximum share of free allocation was defined as the historic share[22] of emissions of those installations in the overall EU ETS cap.[23] A cross-sectoral correction factor was therefore foreseen between 2013 and 2020 to ratchet down free allocations to all operators to the same extent to ensure that the pre-determined limit of free allocation is not exceeded.

The legislation foresaw that the benchmark values were to be multiplied by production values. Operators had a choice of base year for production values, which resulted in a significant inflation of allocation and triggered the application of a significant cross-sectoral correction factor for the period from 2013–2020. In 2013, the cross-sectoral correction factor was around 6% increasing to around 18% by 2020, some 12% on average. This ratchet applied equally to the free allocations of all operators, regardless of the differing extent of exposure to global competition.

The application of the correction factor generated support for a more targeted carbon leakage system for the post-2020 period. As a result, the updates of the carbon leakage list and of the benchmark values should reduce the need, if not completely avoid, the application of the cross-sectoral correction factor. However, in case the correction factor would still be applied, up to 3% of the total quantity of allowances by 2030 will be used to give additional free allocation rather than be auctioned by Member States.[24]

4.5.5 The New Entrants Reserve

For new entrants in the period from 2013–2020, 5% of the total quantity of allowances[25] has been set aside for new investments, either in terms of entirely new installations or significant capacity extensions of existing installations. Harmonised benchmarking rules are set out for allocations to new entrants and the allocations are reduced by the Linear Reduction Factor. The New Entrants Reserve for the period from 2021–2030 is drawn from some 325 million unallocated allowances from the period from 2013–2020, including 200 million allowances that would otherwise have been placed in the Market Stability Reserve.

A large proportion of the new entrants reserve in the period from 2013–2020 was used for the "NER300" fund that provided specific support for demonstration activities for innovative renewable energy technologies. Regarding installations that close, the legislation provides that no free allocation is given any longer to an installation that has ceased its operations unless the operator shows that production will be resumed within a reasonable time. The same rule applies for the partial closure of installations or significant reductions of capacity.

4.5.6 State aid to compensate for the cost of carbon passed through in electricity prices

The Directive's general rule that no free allocation is given to electricity generation is based on the premise that generators are expected to pass through the costs of carbon in electricity prices. As a result, industry sectors have an increased "indirect" cost of electricity prices. In view of this, the Directive states that Member States may grant State aid, i.e., national subsidies, for the benefit of sectors exposed to a significant risk of carbon leakage due to higher electricity prices. Ten Member States grant such State aid,[26] which must nevertheless comply with the EU's State aid guidelines (requiring, in particular, that it does not give rise to distortions of competition within the EU's internal market). This system will continue until 2030, but with much improved transparency.

> Conclusion: Free allocation shields the manufacturing industry from possible negative impacts on its international competitiveness. From 2021, 43% of allowances issued within the EU ETS are given out for free to manufacturing industry according to objective criteria.

4.6 Fairness and aspects of solidarity

The EU ETS creates a level-playing field for greenhouse gas reductions between EU companies within the EU's internal market, leading to a cost-effective and efficient policy. However, it has had to be designed against the background of a wide *per capita* income disparity of more than a ratio of 1:10 between the EU's Member States. Striking the right balance between efficiency on the one hand and solidarity on the other has been of capital importance.

The EU ETS has a general principle that auction revenues accrue to the Member States where the corresponding emissions are generated. However, a redistributive element has also been created. More precisely, 88% of allowances to be auctioned are distributed amongst Member States on the basis of their historical share of verified emissions, while 10% is distributed amongst certain Member States for the purpose of solidarity and growth. Up to 2020, a further 2% was distributed amongst Member States whose emissions were at least 20% below their Kyoto Protocol base-year emissions in 2005.[27] This provision was specifically designed to benefit those Member States that had undergone substantial economic restructuring after the collapse of Communism in Central and Eastern Europe.

This distributional element was instrumental in mobilising political support from all Member States and has proven to be a valuable tool very similar to the system of free allowances for companies. Distributional elements have also been inserted in the revision of the legislation for the period from

2021–2030. The 10% distribution was kept, but instead of the "Kyoto bonus" of 2%, a Modernisation Fund has been established.

The Modernisation Fund benefits ten EU Member States with lower GDP.[28] It consists of approximately 310 million allowances, which could increase by the mid-2020s by up to 75 million allowances if these are not used for free allocation to industry. The financial resources in the Modernisation Fund will be distributed amongst lower-income Member States according to a key defined in Annex IIa to the EU ETS Directive.[29] At least 70% of the Fund's resources will be used for priority projects in renewable electricity generation, improving energy efficiency (including in transport), buildings, agriculture and waste, energy storage and modernising energy networks. The Modernisation Fund will provide no support to energy generation facilities using solid fossil fuels (subject to a narrow exception for district heating systems in Romania and Bulgaria). Priority projects also include support for a "just transition" in "carbon-dependent" regions towards a low-carbon economy including redeployment, reskilling, education or job-seeking initiatives.

The EU ETS also allows for national derogations from the general rule of auctioning in order to support the modernisation of the electricity sector in certain EU Member States. Eight Member States have made use of the derogation[30] to allocate to electricity generators a number of allowances for free, provided corresponding investments are carried out. The total value of reported investment support during the years 2009–2016 is estimated to have been some €11 billion. About 80% of this was dedicated to upgrading and retrofitting infrastructure, while the rest of the investments were in clean technologies or diversification of supply. From 2021–2030, this possibility has been extended, and up to 700 million allowances are available to be used by Member States for this purpose. Investments carried out under this mechanism will generally be subject to competitive bidding, except for small-scale projects of a value below €12.5 million.

Conclusion: The EU ETS takes into account fairness considerations: a redistribution of auctioned allowances reflects income disparities between Member States; modernisation of the energy sector in lower-income Member States is encouraged through financial support; and lower-income Member States can opt for partial free allocation to the power sector in exchange for low-carbon investments.

4.7 The use of auction revenues for low-carbon innovation and climate policies

Economists start from the assumption that it is best to auction all allowances brought to the market. In reality, however, this is not as easy as assumed and therefore most emissions trading systems start with low levels of auctioning that are then gradually expanded. The EU ETS started in 2005 with a political decision that the large majority of allowances should be given out for free. However, from 2013, over half of all allowances have been auctioned – primarily to the power sector – that as a rule were no longer eligible for free allocation. From 2021 onwards, 57% of allowances will be auctioned.

Emissions trading systems are intended to have price effects that flow through supply chains to the final consumer. There have been several studies on when and to what extent these signals are passed through.[31] There has also been much discussion on whether companies were making additional profits by passing through to consumers the price of allowances that they received for free (so-called "windfall profits"). This was particularly highlighted in relation to the power sector, which explains why, from 2013, no free allocation is given to power generators, except for some investment support in eight Member States, as explained earlier.

An Auctioning Regulation fixes the rules for auctioning in detail.[32] It is based on the principles that operators have full, fair and equitable market access, that the same information is available to everybody at the same time and that the organisation and participation in auctions is cost-efficient. An evolution took place from limited auctions by individual Member States to an EU-wide auctioning process using a common auction platform. The European Energy Exchange (EEX) based in Leipzig has been carrying out the role of the EU ETS common auction platform on behalf of 25 Member States. Germany, the UK and Poland have opted out of the common platform and have parallel auction platforms or arrangements. The common auction platform is the most significant auction process for environmental assets ever designed and implemented. So far, over 1000 auctions have been undertaken.

The EU itself, with the help of the European Investment Bank, has also been involved in the sale of allowances through a provision of the Directive that allowed the market value of up to 300 million allowances to be used for investment in innovative renewable energy technologies and commercial demonstration plants of carbon capture and storage (CCS). This so-called "NER300" fund (referring to the New Entrant Reserve from which the 300 million allowances came) resulted in around €2 billion being raised from the selling of these allowances.[33]

For 2021–2030, an Innovation Fund replaces the NER300 instrument and receives at least 450 million allowances. The proceeds will be used for promoting innovation in CCS technology in renewable energy and in industrial low-carbon processes and technologies. This could increase by a further 50 million allowances by the mid-2020s if these are not used for free allocation to industry. Overall, around €10 billion is expected to be available for the Innovation Fund.[34]

Aside from the *de facto* "earmarking" of revenues at EU level under the NER300 and Innovation Fund, it is important to highlight that most of the money generated from auctions goes to Member States. European legislation states that at least half of auction revenues should be used by Member States to tackle climate change and a Declaration by Heads of State and Government from 2008 emphasises the willingness of Member States to do this.[35] Between 2012 and 2017, Member States received total auction revenue of €20.1 billion and approximately 80% of these revenues were used for specified climate- and energy-related purposes.[36] Germany, for example, uses all its EU ETS revenues for its international and national climate funds, Spain for paying for renewable energy incentives, and France earmarks these revenues for improving the insulation of social housing. In 2018, total auction revenue amounted to €14.1 billion, although future revenues will be impacted by the reduction of volumes to be auctioned as the Market Stability Reserve begins to operate.

Conclusion: More than half of the allowances issued are auctioned according to harmonised market rules. Member States use some 80% of these revenues for climate action. An Innovation Fund has been set up with up to 500 million allowances to promote low-carbon innovation in the private sector.

4.8 The use of international credits

In addition to establishing a price for greenhouse gas emissions in Europe, the EU ETS has also been the main driver for emission reduction projects[37] around the world. It has indeed been the main market for, or importer of, credits from the Kyoto Protocol's project-based mechanisms, the Clean Development Mechanism (CDM) and Joint Implementation (JI) and is estimated to be responsible for the use of approximately 1.6 billion international credits up to 2018.[38] Billions of euros of investments took place through the CDM in sustainable development projects in developing countries, including, for example, in renewable energy investments. The sale of credits to operators with compliance obligations under the EU ETS has significantly facilitated the financing of such offset projects around the world.

The EU initially relied solely upon the UNFCCC to generate and validate the offset credits. Exclusions were made, however, for credits based on projects where the emission reductions were not permanent, as is the case for forestry-based credits or for projects that were considered politically unacceptable, such as for new nuclear power stations. In the light of experience, the EU realised that it needed to improve its qualitative and quantitative conditions[39] for the use of international credits.

First, the experience showed that the UNFCCC structure was unable to enforce high environmental standards, and the EU always had qualitative conditions for credits to be used under the EU ETS.[40] Additional quality standards were put in place in 2011 for projects involving HFC-23 destruction and Adipic acid production.[41] In the end, it was found necessary to prevent Joint Implementation credits that lacked credibility from being used for compliance purposes in the EU ETS.[42]

Second, the influx of international credits also had a downside from a quantitative perspective. Exactly at the moment that the economic crisis was creating a surplus of allowances, a large amount of international credits entered the EU ETS. As is explained in Section 4.3, this influx of international credits further inflated the recession-induced surplus and led to the lower-than-expected carbon price in Europe for several years. It was vital that there was a quantitative limit on the overall quantity of credits that could enter the EU ETS.

In 2014, the EU decided to define its 2030 target only in terms of a 40% domestic reduction commitment. This was consistent with the set-up of the Paris Agreement. The approach of the Kyoto Protocol has not been continued and any new market for international credits will come from Article 6 of the Paris Agreement. Article 6 still needs to be operationalised in a way that rewards action going beyond national commitments and "business as usual." The EU ETS will no longer be a source of demand for credits as long as the current legislation is not modified.

Conclusion: The EU ETS absorbed some 1.5 billion tonnes of international credits (CDM and JI). The EU was the main source of demand for such offset credits, which stimulated interest in market-based approaches in many countries. The EU decided on a reductions target for 2030 to be achieved domestically.

4.9 The prospect for international cooperation on carbon markets

At the time of the 1997 Kyoto Protocol, the international carbon market was conceived as something that would happen between the Parties in a "top-down" fashion. In the intervening 20 years it became clear that international carbon pricing was rather the result of choices by national and regional Governments to put obligations on certain economic actors in their jurisdictions rather than being driven by developments at the UN level. Businesses are more optimal actors in a carbon market than governments, as businesses know much better the real costs of abatement measures and technologies.

The EU has led the way in developing actual carbon pricing through its emissions trading system. The EU ETS started in 2005, prior to the first commitment period of the Kyoto Protocol, although its design was completely compatible with UN requirements. While the UNFCCC process remains important in a general climate policy context, there has never been a serious attempt to establish a company-based emission trading system through UN institutions. The Paris Agreement does not even mention carbon pricing explicitly,[43] despite its importance for tackling climate change.

Over time, more countries have put in place carbon pricing. Romania and Bulgaria joined the EU ETS upon their accession to the EU in 2007, and Croatia joined from 2013 when it too became an EU Member State. The first formal linking of the EU ETS with States that were not Member States of the European Union was the extension to the neighbouring countries of the European Economic Area – Norway, Iceland and Liechtenstein – in 2008. Furthermore, the EU is also linking its emissions trading system with Switzerland's system, following agreement that was reached in the margins of the Paris Conference in December 2015. This is the first formal Treaty for linking carbon markets, and the adoption of the ratification Decision by the EU took place in 2017. The linking agreement will enter into force following ratification by both sides.

This broader action, strongly encouraged by the Paris Agreement, is essential. There is a growing recognition that the magnitude of the climate change challenge requires that a price be put on carbon, to create incentives for companies to invest in low-carbon activities. Many developed and developing countries are developing plans to establish their own emissions trading systems.

This is not to underestimate the political challenges to develop legislation to put a price on emissions, whether in the form of emissions trading or taxes on emissions. This has been amply experienced, for example through

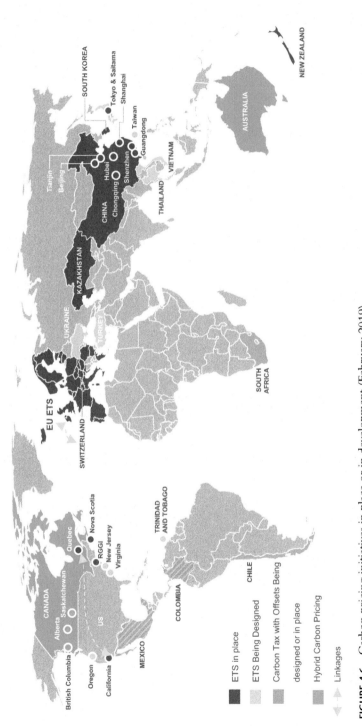

FIGURE 4.6 Carbon pricing initiatives in place or in development (February 2019).

Source: International Emissions Trading Association (IETA) and International Carbon Action Partnership (ICAP)[45]

the United States' inability to pass federal legislation to put a price on emissions over the last 20 years, despite a number of near successes.[44] Australia established a national emissions trading system that would have been linked with the EU ETS, but a change in government halted their national system. Some of the most promising policy developments outside Europe today are happening in Asia. South Korea has a national greenhouse gas emission trading system that was up and running as of January 2015. China has established seven pilot emissions trading systems, to be expanded into a nation-wide system by 2020.[46] Experience of the Kyoto Protocol's Clean Development Mechanism played an important role in enhancing understanding of market-based approaches.

New Zealand also has an emissions trading system that has been in operation since 2008. In the United States, despite the failures at federal level, the northeastern states have been operating the Regional Greenhouse Gas Initiative (RGGI) since 2009,[47] and California has an emissions trading system operating since January 2013. Canada is developing a national carbon pricing system, building on the actions by Quebec, Nova Scotia and other provinces and complementing it with carbon taxes elsewhere.

These national and sub-national experiences offer prospects for developing an international carbon market. In the future, allowances could be traded across jurisdictions and a common carbon price could emerge within a wider geographical area. Just as the EU ETS has provisions for international credits, it also explicitly allows for the linking of carbon markets by means of "mutual recognition" of carbon allowances,[48] through bilateral agreements between the EU and third countries or regions. The possibility of linking also extends to sub-national systems if this were considered desirable.

The "bottom-up" development of an international carbon market via national legislation, as well as the linking agreements between them, inevitably takes time. Rules for monitoring, reporting and verification of emissions are a key infrastructure requirement for a carbon market to function properly. Similar to contemporary financial systems underpinned by robust accounting systems, carbon markets need a solid underpinning by an emissions accounting system. The EU ETS installed a system of self-reporting by companies that is verified by independent third-party verifiers and this approach is being widely followed. One of the main lessons learnt from the EU ETS's pilot-phase was the need for reliable industry-wide and plant-specific emissions data. There needs to be confidence that emissions are reliably measured with high accuracy, which comes from a well-developed set of rules enforced by competent authorities.

In order to facilitate such a process, bilateral cooperation as well as programmes such as the World Bank's Partnership for Market Readiness and the International Carbon Action Partnership has proven to be very valuable.[49] The potential for working together and for more cost-effectiveness is there, but linking systems first needs robust and functioning cap-and-trade systems. With the launch of its national emissions trading system, China has become, along with the EU, a source of inspiration for other economies.[50] They use the carbon market as an instrument to move from making commitments to delivering actual emission reductions, and their "learning by doing" will be invaluable to others.

Conclusion: The EU and other countries such as China and Korea are putting carbon pricing in place. The EU ETS can be linked with other comparable systems, subject to adequate standards and safeguards. Such linking would facilitate the emergence of an international carbon price.

Conclusion

The EU Emissions Trading System is of strategic importance to the EU because international commitments can be achieved cost-effectively and without adverse impacts on the competitiveness of European businesses. During the first 12 years of its existence, emissions covered by the system decreased by 26%. The EU ETS also demonstrated that it was able to promote new jobs, expanding sectors concerned with the energy transition while preparing the EU economy for greater carbon constraint in the future.

The EU ETS has been innovative but is also constantly evolving, as "learning-by-doing" and external economic conditions have developed rapidly. In this respect, two key elements were of major importance: the nature and strictness of the overall emissions cap and the extent of the auctioning of allowances.

The EU has an absolute cap determining the overall environmental ambition of the system. The cap and a declining trajectory of emissions from 2013 was set in a transparent manner before the start of every period. The political climate surrounding the cap decision has always been characterised by a general fear from industry that the cap would be set in too strict a manner. The EU experience illustrated how initially the cap-setting and free allocation has a tendency to be overly generous. Once familiarity with the trading system and trust in the monitoring of emissions is established, the conditions are then set for a gradual tightening of the allocation conditions. The EU did not make the mistake of choosing an output-based cap, based on production levels, that gives no guaranteed environmental outcome while making the system overly complicated in its functioning.

The other interesting evolution is the extent to which the EU ETS started as a system where allocations were almost all made for free, to one where over half of allocation is by auctioning. This transition, coupled with other reforms of the EU ETS, have resulted in the raising of significant amounts of revenue, as mentioned in Section 4.7 earlier. This has not only been an implementation of the "polluter pays" principle but has also created a "double-dividend" whereby revenues generated are deployed to help innovation and deployment of new technologies, for example, or to modernise energy systems and infrastructure, as well as finance capacity building in developing countries. Not only is pollution discouraged by the price of carbon, but the revenues generated by the EU ETS have also been used by Member States and the European Union for further enhancing climate action. Other than environmental taxes, few other policy tools have this "double-dividend" potential by virtue of the successful application of the instrument.

The economic recession in Europe from 2009–2013 served to "road-test" the EU ETS more robustly than expected at the beginning of Phase-2 in 2008. It demonstrated very clearly the way the EU ETS adjusted to the economic cycle, effectively lessening the cost burden on businesses during economically hard times. At the same time, the system was adapted in several ways: through adding the Market Stability Reserve, the strengthening of the Linear Reduction Factor, the continuation of free allocation to parts of industry on the basis of ambitious benchmarks or the extension of the Innovation Fund to industrial processes. Subtle redistributive elements have also ensured the support of stakeholders and governments in less wealthy Member States, who also understand that if the costs of abatement are lower in these countries then revenue flows will tend to gravitate towards them.

To summarise, the last ten years and particularly the recent changes made in readiness for Phase-4 of the EU ETS (2021–2030) have seen a robust adaptation of the EU ETS that won wide support, including in the Council and European Parliament. Adjustments have enhanced the impact of the "double-dividend" in such a way as to strengthen fairness elements and the safeguards to maintain the competitiveness of European businesses, while maintaining the effectiveness of the environmental instrument itself.

Notes

1 Delbeke, Jos and Vis, Peter (eds.) (2015). *EU Climate Policy Explained*. London: Routledge.
2 Delbeke, Jos (ed.) (2006). *EU Energy Law, Volume IV; EU Environmental Law: The EU Greenhouse Gas Emissions Trading Scheme*. Claeys & Casteels.
3 Between 2005 and 2007, the penalty rate was €40 per tonne of excess emissions plus the obligation to make up the shortfall.
4 See for more information, "Trends and Projections in the EU ETS in 2018". EEA Report No. 14/2018.
5 Ellerman, A.D. & Buchner, B.K. "Over-Allocation or Abatement? A Preliminary Analysis of the EU ETS Based on the 2005–06 Emissions Data", Environmental and Resource Economics, October 2008, Volume 41, Issue 2, pp 267–287 https://doi.org/10.1007/s10640-008-9191-2 ; Ellerman, D., Convery, F., Perthuis, C. (2010). "Pricing Carbon: The European Union Emissions Trading Scheme", Cambridge University Press.
6 Increase of CO_2 emissions in years 2014, 2015, 2016 and 2017 compared to 2013. See Section 4 of the 2017 Carbon Market Report (COM(2017)693 of 23/11/2017) (use the 2018 Carbon Market Report COM(2018)842 final, at https://ec.europa.eu/clima/sites/clima/files/ets/docs/com_2018_842_final_en.pdf?) in conjunction with Section 2.1 of Progress Report prior to COP-24 in Katowice (COM(2018)716 of 26.10.2018).
7 European Commission website article dated 04/06/2019: https://ec.europa.eu/clima/news/emissions-trading-emissions-have-decreased-39-2018_en.
8 Commission Regulation (EU) No 176/2014 of 25 February 2014 amending Regulation (EU) No 1031/2010 in particular to determine the volumes of greenhouse gas emission allowances to be auctioned in 2013–20; OJ L 56, 26.2.2014, pp. 11–13. http://eur-lex.europa.eu/legal-content/EN/TXT/PDF/?uri=CELEX:32014R0176&from=EN.
9 Decision (EU) 2015/1814 of the European Parliament and of the Council concerning the establishment and operation of a market stability reserve for the Union greenhouse gas emission trading scheme and amending Directive 2003/87/EC. http://eur-lex.europa.eu/legal-content/EN/TXT/PDF/?uri=CELEX:52014PC0020&from=EN.
10 Commission Decision of 27 April 2011 determining transitional Union-wide rules for harmonised free allocation of emission allowances pursuant to Article 10a of Directive 2003/87/EC of the European Parliament and of the Council, 2011/278/EU; OJ L 130, 17.5.2011, pp. 1–45. http://eur-lex.europa.eu/legal-content/EN/TXT/PDF/?uri=CELEX:32011D0278&from=EN.
11 Although the two fallback benchmarks are based on inputs.
12 Case C-460/15 Schaefer Kalk GmbH & Co KG; Case C-456/15 Borealis Polyolefine GmbH (with other Joined Cases); Case C-180/15 Borealis AB and others; Case C-506/14 Yara Suomi Oy and others. For more details see "Report on the Functioning of the European Carbon Market" (COM(2017)693 of 23.11.2017 and "Report on the Functioning of the European Carbon Market" COM(2018)842 final. https://ec.europa.eu/commission/sites/beta-political/files/report-functioning-carbon-market_en.pdf and https://ec.europa.eu/clima/sites/clima/files/ets/docs/com_2018_842_final_en.pdf). Court cases are listed in each document in Appendix 5.

13 Commission Decision of 24 December 2009 determining, pursuant to Directive 2003/87/EC of the European Parliament and of the Council, a list of sectors and subsectors which are deemed to be exposed to a significant risk of carbon leakage, 2010/2/EU; OJ L 1, 5.1.2010, pp. 10–18. http://eur-lex.europa.eu/legal-content/EN/TXT/PDF/?uri=CELEX:32010D0002&from=EN. This decision was subsequently amended in 2011, 2012 and 2013; for more details see: http://ec.europa.eu/clima/policies/ets/cap/leakage/documentation_en.htm.

14 2014/746/EU: Commission Decision of 27 October 2014 determining, pursuant to Directive 2003/87/EC of the European Parliament and of the Council, a list of sectors and subsectors which are deemed to be exposed to a significant risk of carbon leakage, for the period from 2015 to 2019 (notified under document C(2014)7809); OJ L 308, 29.10.2014, pp. 114–124. http://eur-lex.europa.eu/legal-content/EN/TXT/PDF/?uri=CELEX:32014D0746&from=EN. The second list has been extended to apply for the year 2020 by Article 4 of Directive (EU) 2018/410.

15 The level of disaggregation for sectors and sub-sectors was undertaken at a detailed level, so-called "NACE-4," with more disaggregated analysis for specific sub-sectors where this was considered justified.

16 European Commission: Staff Working Document, *Impact Assessment: Accompanying the Proposal for a Directive of the European Parliament and of the Council amending Directive 2003/87/EC to enhance cost-effective emission reductions and low-carbon investments*, reference SWD(2015) 135 final of 15.07.2015. https://ec.europa.eu/clima/sites/clima/files/ets/revision/docs/impact_assessment_en.pdf.

17 The generally applicable criterion is the mathematical produce of emissions and trade intensity higher than 0,2.

18 Article 10b of Directive 2003/87/EC, as amended by Directive (EU) 2018/410, includes supplemental routes for inclusion.

19 Commission Delegated Regulation (EU) 2019/331 of 19.12.2018, determining transitional Union-wide rules for harmonised free allocation of emission allowances pursuant to Article 10a of Directive 2003/87/EC of the European Parliament and of the Council, published in the Official Journal of the European Union, L 59, 27.2.2019, pp. 8–69. https://eur-lex.europa.eu/legal-content/EN/TXT/PDF/?uri=OJ:L:2019:059:FULL&from=EN.

20 Article 10a(2)) and (21), and recital (12).

21 Article 10a(5) of Directive 2003/87/EC.

22 The "historic share" is established by emissions levels in the years 2005–2007.

23 From 2021, the primary legislation defines the maximum share of free allocation in a more upfront and transparent manner.

24 If this 3% is not needed to avoid a correction factor applying, some of the allowances will be used to increase the size of the Innovation Fund and the Modernisation Fund.

25 Three hundred million allowances were taken from this New Entrants Reserve to fund the "NER300" fund for innovative renewable energy technologies and carbon capture and storage projects (though none of the latter were actually funded due to a lack of projects).

26 See Section 4.1.3 of the 2018 Carbon Market Report COM(2018)842 final. https://ec.europa.eu/clima/sites/clima/files/ets/docs/com_2018_842_final_en.pdf.

27 Article 1(12) (inserting new Article 10(2) and new Annexes IIa & IIb) of Directive 2009/29/EC of the European Parliament and of the Council of 23.4.2009 amending Directive 2003/87/EC so as to improve and extend the greenhouse gas emission allowance trading scheme of the Community; OJ L 140, 5.6.2009,

pp. 63–87. http://eur-lex.europa.eu/legal-content/EN/TXT/PDF/?uri=CELE X:32009L0029&from=EN.

28 Member States with less than 60% of EU average per capita GDP in 2013, at market prices: Bulgaria, Czech Republic, Estonia, Croatia, Latvia, Lithuania, Hungary, Poland, Romania and Slovakia.

29 The distribution of funds is based on both verified emissions and GDP of beneficiary Member States, as set out in the 2014 European Council Conclusions.

30 Article 10c of the Directive, made use of by Bulgaria, Cyprus, Czech Republic, Estonia, Hungary, Lithuania, Poland and Romania.

31 Sijm, J., Hers, S., Lise, W., Wetzelaer, B., (2008) "*The Impact of the EU ETS on Electricity Prices*" Final report to the European Commission (DG Environment), Energy research Centre of the Netherlands, Petten/Amsterdam; Alexeeva-Talebi, V. (2010), "*Cost Pass-Through in Strategic Oligopoly: Sectoral Evidence for the EU ETS*", ZEW Discussion Paper 10-056, Mannheim; Lise, W., Sijm, J. & Hobbs, B. (2010). "*The impact of the EU ETS on prices, profits and emissions in the power sector: Simulation results with the COMPETES EU20 model.*" Environmental and Resource Economics. Vol. 47. pp. 23-44. 10.1007/s10640-010-9362-9; Sijm, J., Chen Y., and Hobbs, B.F., (2012) "*The impact of power market structure on the pass-through of CO$_2$ emissions trading costs to electricity prices – A theoretical approach*". https://www.sciencedirect.com/science/article/pii/S0140988311002301?via%3Dihub; Jouvet, P-A., Solier, B., (2013) "*An overview of CO$_2$ cost pass-through to electricity prices in Europe*", Energy Policy, Volume 61, 2013, pp. 1370-1376; De Bruyn S. M. et al (2015) "*Ex-post investigation of cost pass-through in the EU ETS: An analysis for six sectors*", analysis undertaken on behalf of the European Commission, DG Climate Action (views expressed are those of the authors): https://ec.europa.eu/clima/sites/clima/files/ets/revision/docs/cost_pass_through_en.pdf.

32 Commission Regulation (EU) No 1031/2010 of 12 November 2010 on the timing, administration and other aspects of auctioning of greenhouse gas emission allowances pursuant to Directive 2003/87/EC of the European Parliament and of the Council establishing a scheme for greenhouse gas emission allowances trading within the Community (Official Journal of the European Union, L 302, 18.11.2010).

33 More details of about this Fund, and the investments it has contributed to, are on the website of the European Commission. http://ec.europa.eu/clima/policies/lowcarbon/ner300/index_en.htm.

34 European Commission: Press Release of 26 February 2019 *Towards a climate-neutral Europe: EU invests over €10bn in innovative clean technologies*, reference IP/19/1381. https://europa.eu/rapid/press-release_IP-19-1381_en.htm.

35 See Council of European Union document date 12/12/2008 (Reference: 17215/08) "Energy and Climate Change – Elements of the Final Compromise": "The European Council Recalls That Member States Will Determine, in Accordance with Their Respective Constitutional and Budgetary Requirements, the Use of Revenues Generated from the Auctioning of Allowances in the EU Emissions Trading System. It Takes Note of Their Willingness to Use at Least Half of This Amount for Actions to Reduce Greenhouse Gas Emissions, Mitigate and Adapt to Climate Change, for Measures to Avoid Deforestation, to develop Renewable Energies, Energy Efficiency as Well as Other Technologies Contributing to the Transition to a Safe and Sustainable Low-Carbon Economy, Including Through Capacity Building, Technology Transfers, Research and Development." www.consilium.europa.eu/uedocs/cms_data/docs/pressdata/en/ec/104672.pdf.

Damien Meadows, et al.

Analysis of the use of auction revenues by Member States is available at https://
ec.europa.eu/clima/sites/clima/files/ets/auctioning/docs/auction_revenues_
report_2017_en.pdf.

37 Although it is important to keep in mind that the credits of these emissions reductions outside Europe give rise to a corresponding increase in emissions within the
EU. This feature has been frequently misunderstood.

38 See Section 4.1.3 of the 2018 Carbon Market Report COM(2018)842 final. https://
ec.europa.eu/clima/sites/clima/files/ets/docs/com_2018_842_final_en.pdf.

39 The study on additionality is available at: https://ec.europa.eu/clima/sites/clima/
files/ets/docs/clean_dev_mechanism_en.pdf.

40 The initial conditions excluded forestry credits, see Directive 2004/101/EC for
more details. In 2008, it was decided that the only new CDM projects for which
the EU ETS should provide demand post-2013 should be those established in
least developed countries, see Directive 2009/29/EC.

41 Commission Regulation (EU) No 550/2011 of 7 June 2011 on determining,
pursuant to Directive 2003/87/EC of the European Parliament and of the Council, certain restrictions applicable to the use of international credits from projects
involving industrial gases; OJ L 149, 8.6.2011, pp. 1–3. http://eur-lex.europa.eu/
legal-content/EN/TXT/PDF/?uri=CELEX:32011R0550&from=EN.

42 See Article 58(2) of Commission Regulation (EU) No 389/2013 of 2 May 2013
establishing a Union Registry pursuant to Directive 2003/87/EC of the European
Parliament and of the Council, Decisions No 280/2004/EC and No 406/2009/
EC of the European Parliament and of the Council. https://eur-lex.europa.eu/
legal-content/EN/TXT/?uri=CELEX:32013R0389.

43 Paragraph 137 of the accompanying COP Decision "also recognises the important role of providing incentives for emissions reduction activities, including tools
such a domestic policies and carbon pricing."

44 The McCain-Lieberman bill for a comprehensive cap and trade system was first
proposed in 2003. In 2009, the United States came close to having a similar
system to the EU ETS, by the House of Representatives' passage of the Waxman-
Markey bill. However, the companion bill in the Senate, the Kerry-Boxer bill
(S. 1733), was not passed, and near-term prospects for passage of Federal legislation that puts a price on greenhouse gas emissions now appear low.

45 More information on the state of play can be found at https://icapcarbonaction.
com/en/icap-status-report-2019. An updated interactive map (copyright ICAP)
is available at www.icapcarbonaction.com/ets-map.

46 Han, G., Olsson, M., Hallding, K., and Lunsford, D. (2012). "China's Carbon Emission Trading: An Overview of Current Development". FORES, Bellmansga-
tan 10, SE-118 20 Stockholm. www.sei-international.org/mediamanager/docu
ments/Publications/china-cluster/SEI-FORES-2012-China-Carbon-Emissions.pdf.

47 ICAP (2015). "Emissions Trading Worldwide: ICAP Status Report 2015". https://icap
carbonaction.com/images/StatusReport2015/ICAP_Report_2015_02_10_
online_version.pdf.

48 Article 25.

49 See, for example: Bruyninckx, H., Qi, Y., Nguyen, Q.T., and Belis, D. (Eds.), (2014)
*The Governance of Climate Relations between Europe and Asia: Evidence from China
and Vietnam as Key Emerging Economies*. Cheltenham and Northampton, MA:
Edward Elgar, 2014. ISBN 978-1781955987.

50 See, for example, Wettestad, J., and Jevnaker, T. "Rescuing EU Emissions Trading: The Climate Policy Flagship". https://webgate.ec.europa.eu/ilp/pages/saml-
request.jsf.

5

THE EFFORT SHARING REGULATION

Artur Runge-Metzger and Tom Van Ierland

Introduction

Despite the wide support for climate action in Europe, there is also a genuine concern about the socioeconomic impacts of the transition to a low-carbon economy. For that reason, avoiding higher transition costs than necessary for European businesses and consumers is of importance. Economists have always underlined the importance of reaching climate objectives at the lowest cost possible.

In addition to cost-effectiveness comes fairness, which has many dimensions too. The EU accepts its responsibility as a group of developed countries, whose high level of development puts the onus on it to reduce greenhouse gas emissions at a faster pace than other countries. Furthermore, there is also the dimension to address a fair distribution of effort between the EU's Member States. Political decisions on the EU's overall climate ambition, as well as on the individual greenhouse gas emission reduction targets of each Member State, have always been agreed unanimously. This requires great attention to the political, economic and industrial differences between Member States and the distributive impacts of the EU's overall climate policy.

This chapter looks more closely at the details of this differentiation of effort between Member States, more specifically in the sectors of the economy not covered by the EU ETS. Through an equitable sharing of effort, over the past 20 years the EU has been able to demonstrate its continued commitment to global leadership on climate change.

5.1 Emissions from the non-ETS sectors

While the EU ETS regulates the greenhouse gas emissions from large fixed installations, the Effort Sharing Decision[1] and its successor, the Effort Sharing Regulation,[2] regulates emissions of the sectors outside the scope of the EU ETS, the so-called "non-ETS sectors." These emissions currently cover around 60% of the EU's greenhouse gas emissions. They typically come from a set of diffuse sources such as from road vehicles, the heating of private households and business premises, small- and medium-sized industry, agriculture, waste management facilities and products containing fluorinated gases (which are often powerful greenhouse gases). Their diffuse nature makes them less suitable to be readily incorporated into the EU ETS.

Road transport is the largest source of these, representing more than a third of the emissions in the non-ETS sectors, followed by the heating of buildings and agriculture. The non-ETS sectors reduced their emissions

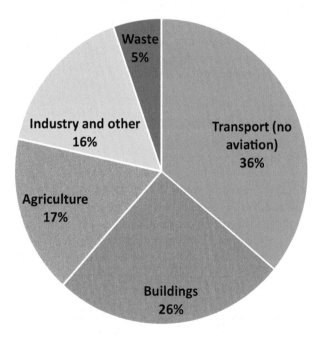

Non ETS emissions in 2017

Waste
5%

Industry and other
16%

Transport (no
aviation)
36%

Agriculture
17%

Buildings
26%

FIGURE 5.1   Main greenhouse gas emitting sectors of the non-ETS in 2017

Source: EEA Trends and Projections report 2018, EEA GHG projections dataset 2018

more slowly than the EU ETS sectors. In the period from 2005–2017, road transport, heating, agriculture, waste and fluorinated gases achieved emission reductions of 11% in contrast to the 26% reduction realised by sectors covered by the EU ETS.

The EU's emissions in the non-ETS sectors were stable during the period from 1990–2005, and they only started reducing significantly from 2007. The main reason for this lack of progress was the road transport emissions, which by 2007 had risen by 30% compared to their level in 1990. Emissions of fluorinated gases have grown by 60% since 1990, though this remains in absolute amounts much smaller than road transport's CO_2 emissions. As for the other non-ETS sectors, they have all managed to reduce their emissions, with waste, heating of buildings and agriculture reducing emissions by 42%, 22% and 20% respectively by 2017, compared to 1990. After 2014, though, there has been a rebound in total non-ETS emissions, driven foremost by a reversal in transport emissions.

There are important policies applicable to the non-ETS sectors, such as on road vehicles, buildings and agriculture. However, most of the policies that affect the emissions of these non-ETS sectors are determined by the Member States, such as national taxation policies, urban planning, transport and mobility policies, as well as the granting of environmental permits. Crucial for success in reducing greenhouse gas emissions is ensuring the coherence of policies at the respective levels of public intervention, from the European to the national or local level. If all pull together in the same direction, impressive results can be achieved. If there are inconsistencies, on the other hand, such as where company-car taxation favours cars and creates disincentives to use public transport, the combination of policies will be much less efficient. Mobilising policy levers at the right level of governance in a way consistent between different levels of governance is absolutely key to reduce emissions in these non-ETS sectors.

Of course, all EU Member States have signed up to and ratified the UN Framework Convention on Climate Change, the Kyoto Protocol and the Paris Agreement. It is for the Member States to take action themselves and to ensure the consistency of actions at different levels of governance. The European Union, for its part, endeavours to concentrate its efforts and regulate on those areas where there is a clear added value. Apart from the EU ETS, an important example in this respect is vehicle efficiency legislation: not all Member States manufacture cars, yet cars are widely sold and used across the EU. As a consequence, there is strong logic for the EU to regulate the emissions performance of cars, whereas the promotion of public transport or of cycling in urban areas, for example, is more coherently managed at the local level.

Conclusion: Non-ETS sectors give rise to emissions from a variety of sectors unsuitable for inclusion into EU ETS. Action at EU level must be accompanied by national and municipal policies and measures. Reductions in emissions have proven more difficult to realise, and a significant reinforcement of these will be required in the future.

5.2 Effort Sharing 2013–2020

In view of a coherent policy framework at European level, Member States adopted in 2002 a so-called "Burden Sharing Agreement"[3] for the period from 2008–2012. It shared their joint commitment taken under the Kyoto Protocol, covering all emissions of the economy. Furthermore, these "Burden Sharing" targets for Member States could also be met through offset credits from the Clean Development Mechanism.[4] However, in view of the creation of the EU ETS and its uniform carbon price applicable across a large share of the emissions of each Member State, it was not straightforward to continue this method of sharing national targets for all emissions, including those of the EU ETS, beyond the first commitment period of the Kyoto Protocol.

5.2.1 Setting differentiated targets

That is why in 2008 the Commission proposed the EU's "Effort Sharing Decision,"[5] which "extracted" the EU ETS from targets set for Member States, leaving the differentiated national targets only covering the non-ETS sectors for the period from 2013–2020. Targets in 2013 were determined as being what a Member State's average emissions were in the years 2008, 2009 and 2010, while 2020 targets were expressed as a percentage change compared to 2005.[6] Finally, intermediate targets were defined as a linear extrapolation between the 2013 and 2020 targets.

The concern to ensure fairness was at the heart of the 2020 target-setting exercise. Member States with high income levels, expressed through their relative ranking above the EU average in GDP *per capita* terms in 2005,[7] were required to reduce emissions compared to 2005 by a maximum of 20%. Conversely, Member States with lower *per capita* GDP where allowed to reduce emissions by less than the EU average, i.e., almost 10% below 2005 levels. Thirteen Member States were allowed to increase their emissions by 2020 compared to 2005, up to a maximum of plus 20%. That maximum applied for Bulgaria, which had, and still has, the lowest *per capita* income of any EU Member State.

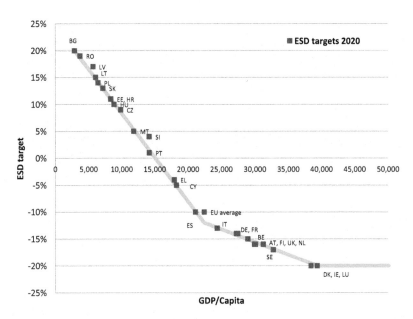

FIGURE 5.2 2020 national Effort Sharing targets compared to 2005, resulting from the methodology in relation to 2005 GDP *per capita*

Source: European Commission calculations based on Eurostat and reported GHG emissions taken from Impact Assessment of Effort Sharing Regulation[8]

For an overview of the distribution of the 2020 non-ETS Member States' targets, see Figure 5.2.

5.2.2 Developing more elements of redistribution

This differentiation allowed the comparatively lower income Member States to increase emissions in sectors where consumption levels and associated emission levels were typically still well below the EU average, such as in transport. Overall, the 2020 climate and energy package had several of these redistributional elements included. Not only were the non-ETS greenhouse gas reduction targets differentiated by taking into account different income levels, but so were the 2020 renewable energy targets for each Member State.

However, the lower income Member States, which typically had a more carbon inefficient industrial structure, were covered by the EU ETS that effectively harmonised effort for all participants irrespective of where they were located (see Chapter 4). For example, a steel plant in a lower income

Member State was subject to the same free allocation rules and stringency, as well as the same carbon price, as its counterparts in higher income Member States. Although the EU ETS created a level playing field for all industries, there was provision made for the redistribution of auctioning revenues towards lower income Member States, allowing them to compensate for the costs of modernisation of their economies.

These redistributional elements were a strong requirement for several Member States to be able to accept the overall architecture of 2020 climate and energy targets. The Impact Assessment accompanying the 2020 climate and energy proposals[9] indicated clearly that a target setting exercise within the EU based on cost-efficiency only would have, in relative terms, higher cost impacts on lower income Member States than higher income ones. By applying the redistributional elements of target differentiation and auctioning revenue redistribution, cost impacts were projected to be distributed much more equally between Member States as a share of GDP *per capita*, as is shown in Figure 5.3.

5.2.3 Experience to date

In 2017, emissions in the non-ETS sectors were around 11% below 2005 levels. This makes the non-ETS sectors already in line with their 2020 target of minus 10% compared to 2005 levels. The non-ETS sectors are expected to maintain this over-compliance through to 2020, but there is no reason to be complacent. Emissions went as low as −14% in 2014, but rebounded in 2015, 2016 and 2017. With significantly lower oil prices and higher economic growth, emissions in the road transport resumed; weather conditions also resulted in increased emissions.

This underlines that solid implementation of existing policies remains of the greatest importance. Transport is a good example. Legislation at the EU level, notably those related to CO_2 standards for passenger cars, improves the efficiency of the car fleet over time. However, Member States have a critical task to complement these efforts with other policies, such as fuel and road pricing policies to manage transport demand. Similarly restructuring of the Common Agricultural Policy, such as tackling the overproduction of certain agriculture produce and productivity gains, have led to reductions in greenhouse gas emissions. Nevertheless, further focused mitigation actions will be needed and in this context Member States have resources available for rural development programmes under the European Agricultural Fund for Rural Development (EAFRD). Finally, EU measures in the context of the Energy Performance for Buildings Directive[10] and Ecodesign measures[11] have significantly improved the insulation of new buildings and the efficiency of newly

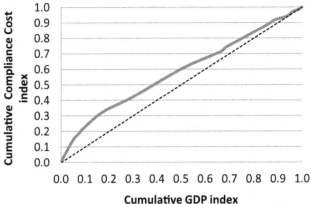

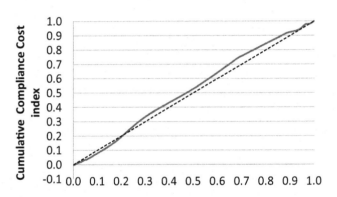

FIGURE 5.3 Distribution of costs of the 2020 climate and energy package, comparing impacts of cost efficient distribution of non-ETS and renewable energy target with redistributed targets and auctioning revenue based on relative income levels

Source: Model-based Analysis of the 2008 EU Policy Package on Climate Change and Renewables[12]

installed boilers. With the relatively low replacement rates of the building stock, however, Member States still have a strong role to play in incentivising energy efficiency improvements and reductions in greenhouse gas emissions for existing buildings.

Ex post evaluation of climate policies indicates that the strongest drivers for emission reductions have been driven by innovation in low-carbon technologies, such as renewable energy, as well as raising productivity and efficiency in the economy. Structural change between economic sectors, such as away from manufacturing towards services, has so far had only a marginal effect on reductions of greenhouse gas emissions across the EU.[13]

5.2.4 Flexible provisions

While the EU is in overall compliance with its 2020 targets, not all individual Member States are (Figure 5.4). A number of Member States with relatively deep reduction targets are expected to have shortfalls for achieving their targets domestically by 2020. Malta, which has a target that allows it to increase emissions by 5% compared to 2005, is expected to have a shortfall.

The Effort Sharing Decision recognises that it may be difficult for all Member States to achieve their targets domestically every single year, due to the inherent variability of emissions, for instance related to heating. Therefore, a number of flexibilities are explicitly allowed. There is flexibility within the period, notably the possibility of "banking" over-compliance in one year to the next and limited "borrowing" from the emission allocations of future years, both within the period until 2020.

In case these "banking" and "borrowing" flexibilities would not be sufficient, "trade" is also allowed, whereby a Member State in shortage can buy part of the over-delivery of another one. This trade is also potentially an incentive to invest in overachieving targets in Member States where reductions can be achieved at lower costs.

For the period from 2012–2020 however, this incentive is not expected to be significant as the EU as a whole is expected to deliver its target. Only a few Member States are expected to miss their target domestically over the whole period. For the three-year 2013–2015 period, 27 Member States achieved their targets. Only Malta's emissions exceeded the target for each of these three years. Malta has therefore already used the flexibility provisions under the Effort Sharing Decision, including the acquisition of tonnes of overachievement from other Member States that were in over-compliance.

Conclusion: In the period from 2013–2020, fairness was primarily achieved through the differentiation of targets for Member States on the basis of GDP *per capita* and the partial redistribution of auctioning revenues under the EU ETS. Flexible arrangements facilitated the delivery of the targets in a cost-effective manner.

See table 4, section 7, annex to the 2018 Progress report SWD(2018) 453 , part 6/6

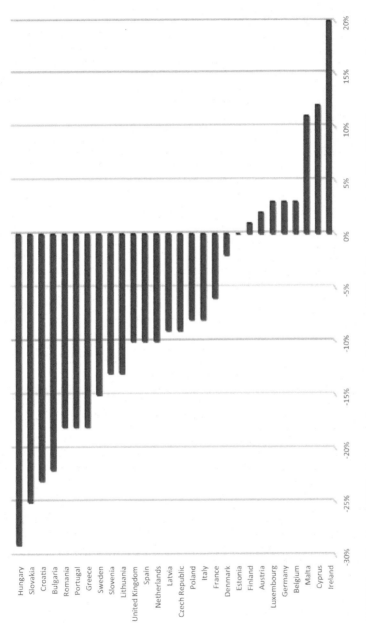

FIGURE 5.4 Projected 2020 relative gap between emissions and Effort Sharing targets (in % of 2005 base year emissions); negative and positive values respectively indicate over-delivery and shortfall

Source: 2018 Progress Report, European Commission[14]

5.3 Differentiation and flexibilities allowed for 2021–2030

As part of its overall strategy, the 2014 European Council endorsed the greenhouse gas emissions reduction target for 2030 of 30% for the non-ETS sectors and the continuation of the approach to differentiate targets between Member States. However, a number of additional elements were added. A correction was suggested for the target differentiation within the group of higher income Member States. Additional flexibility was added related to the EU ETS and the so-called "Land Use, Land Use Change and Forestry" sector, covering emissions and absorption of CO_2 from the different land use sectors, mainly the agriculture and forestry sectors.

5.3.1 Continuation of the differentiated target approach

The methodology of target setting based on GDP *per capita* as followed for 2020 would be continued. Given that the target was increased by 20% from 2020 to 2030, the target range also had to be increased by 20%, resulting in a range for 2030 between 0% to −40% compared to 2005 emission levels.

The target setting for 2030 required an update of the 2005 GDP *per capita* data as some Member States suffered significantly during the economic recession that started from 2008. Therefore, GDP *per capita* data from 2013 were used, and this resulted in a differentiation as represented in the blue (lower) line in Figure 5.5.

For some Member States, this clearly shows the impact of the economic downturn. For instance, using 2005 GDP *per capita*, Spain was very close to the EU average GDP *per capita*, resulting in a target similar to the overall EU ambition level for 2020. However, when setting the 2030 targets, using 2013 GDP *per capita*, the impact of the economic recession resulted in Spain having a GDP *per capita* below the EU average. This therefore resulted in a lower 2030 target for Spain compared to the EU's overall ambition level for 2030. A contrasting example is Germany, which for 2020 received a target equal to France, but which, based on 2013 GDP *per capita*, received a target 1% more ambitious than France for 2030, underlining the stronger economic performance of Germany compared to France over the period from 2005–2013.

5.3.2 More differentiation amongst higher income Member States

The European Council also asked that, within the group of higher income Member States, the targets would reflect relative differences in the

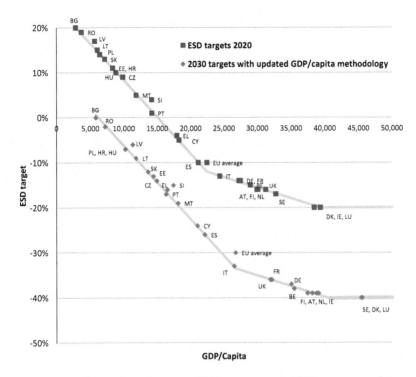

FIGURE 5.5   Comparison between 2020 targets and 2030 targets applying
updated methodology

Source: European Commission calculations based on Eurostat and reported GHG emissions,
taken from Impact Assessment of Effort Sharing Regulation[15]

cost-effectiveness of achieving reductions. Some higher income Member
States did have concerns arising from particularly high marginal abatement
costs within their Member State.

To operationalise this request, different scenarios were constructed comparing projections of potential emission reductions under "cost-efficient and with
existing policies" situations with the proposed targets based on GDP *per capita*.
Figure 5.6 shows the results. A positive number indicates that cost-effective, current policy baseline projections show a gap between emissions and the proposed
targets, while negative numbers indicate that there would be an overachievement
of the respective target by 2030.

What clearly emerged from this analysis was that two higher income Member States, Ireland and Luxembourg, showed a particularly large gap of 15%
or more in all scenarios of what they were likely to achieve cost-effectively

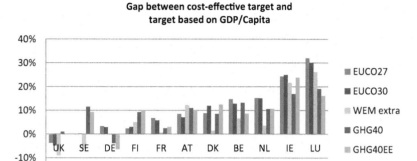

FIGURE 5.6 Gap between GDP-based 2030 targets and cost-efficient emission reductions for high income Member States (as a % of 2005 emissions)

Source: Commission calculations based on PRIMES, GAINS, Eurostat and EEA 2015

and with existing policies. A second group of higher income Member States, namely Austria, Denmark, Belgium and the Netherlands also tended to have a large gap, although below 15%. Three Member States, UK, Germany and France showed no gap or a small gap of around 5% or less. Finally, two higher income Member States, Sweden and Finland, had a gap that varied depending on which set of scenarios was used.

These significant differences were acknowledged and targets where therefore adjusted for Ireland and Luxembourg, reducing them by 9%. Targets for Austria, Denmark, Belgium and the Netherlands where also reduced, albeit by a much smaller amount of 3%. For Sweden and Finland, no target adjustment was made, while for the UK, Germany and France the target was increased by 1%. Overall, this resulted in a basically unchanged target in terms of overall ambition level within this group of higher income Member States. However, the refined differentiation recognised that some Member States had a bigger challenge than others to achieve a target based purely on a GDP *per capita* basis.

5.3.3 Towards the convergence of per capita emissions by 2030

The negotiation on the precise differentiation of the targets resulted in a balanced outcome leading to the 30% target compared to 2005 for the EU as a whole. These 2030 targets take into account the differences in capacity to act between the diverse EU Member States.

TABLE 5.1 2030 targets compared to 2005 emission levels for non-ETS sectors per Member State

LU	−40%	NL	−36%	MT	−19%	LT	−9%
SE	−40%	AT	−36%	PT	−17%	PL	−7%
DK	−39%	BE	−35%	EL	−16%	HR	−7%
FI	−39%	IT	−33%	SI	−15%	HU	−7%
DE	−38%	IE	−30%	CZ	−14%	LV	−6%
FR	−37%	ES	−26%	EE	−13%	RO	−2%
UK	−37%	CY	−24%	SK	−12%	BG	0%

Source: Annex 1 of Effort Sharing Regulation (EU) 2018/842

Table 5.1 gives an overview of what was finally adopted in the Effort Sharing Regulation.[16]

Crucial to note is that analysis shows the final outcome also leads to a stronger convergence between the levels of allowed *per capita* emissions by 2030 compared to 2020. By 2030, 21 Member States are projected to have an allocated emissions level *per capita* within a range of one tonne above or below the EU average (see lower part in Figure 5.7). Of course, the eventual emissions *per capita* in 2030 will depend on effective reductions achieved and the extent to which Member States employ the flexibilities that are allowed (e.g., transfers, as explained earlier).

One can nevertheless conclude that the discussion on differentiated target setting was most worthwhile as it delivered not only a political agreement but also fairness over time by opening a path towards significant convergence in *per capita* emissions within the EU.

5.3.4 The starting point

Compliance targets are actually set for every year, expressed as Annual Emissions Allocations,[17] starting in the year 2021. Annual Emissions Allocations in the period between 2021 and 2030 are then defined, by a linear interpolation, on a straight line between the starting point and the 2030 end point. The starting level denominated in quantities of tonnes of CO_2-equivalent is the average of 2016, 2017 and 2018 emissions in the non-ETS, these being the most recent emissions known in the year 2020 when the absolute amounts of Annual Emissions Allocations will be determined. It was decided that the precise starting point of the trajectory over time would be between 2019 and 2020.[18]

This starting point was a difficult compromise between two positions as represented in a stylised fashion in Figure 5.8.

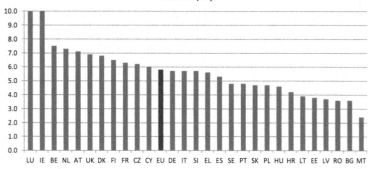

2005 GHG/Cap

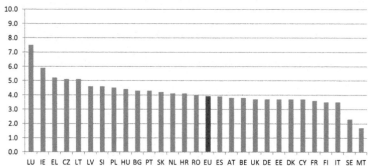

2030 GHG/Capita allocations

FIGURE 5.7 *Per capita* emissions in the non-ETS sectors in 2005 and projected *per capita* allowed emissions in the non-ETS sectors by 2030

Source: Commission calculations based on PRIMES, and GAINS.[19] For 2005, LU and IE emissions *per capita* are 21.5 and 11.6 tonnes *per capita* respectively.

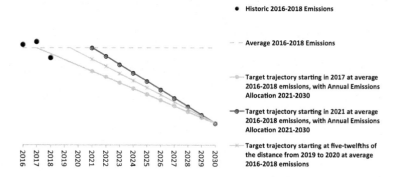

FIGURE 5.8 Stylised representation of possible extreme options to set the starting point in 2021 for the Annual Emissions Allocation and the subsequent linear target trajectory to the 2030 target

Source: Stylised graphic by co-author

One school argued that as climate policies take time to have their full effect, emissions will continue to decrease below their average level in the years 2016–2018. If the starting point would be set in 2020 at the average of 2016–2018 emission levels as represented by the red line, then one would expect targets in the early years of the decades to be overachieved. Member States would then be allowed to build up surpluses in the non-ETS that would reduce the incentive to take further action to effectively achieve their 2030 target. Indeed, through the banking of Annual Emissions Allocations they could deviate from their target later in the period from 2021–2030. Therefore, they proposed to start the target trajectory earlier, for instance in 2017 as represented by the green line in the following figure. This would guarantee a gradual reduction of emissions and reduce the risk for a build-up of surpluses early in the period from 2021–2030.

However, others pointed out that this approach would be flawed if, for some reason, emissions increase or did not decrease sufficiently over the period from 2017–2021. In such a case, the non-ETS sectors would immediately start the period with a deficit in 2021, even in situations where the Member State was fully in-line with its non-ETS targets for the period up to 2020.

Although this would not normally be expected to happen to the EU as a whole – though note that emissions did rebound in 2015, 2016 and 2017 – this situation could arise for a number of Member States individually and confront them with a circumstance whereby they would need to acquire surplus Annual Emissions Allocations early in the period from other Member States. However, these other Member States may either not have been able to build up any surpluses or may not feel confident enough to sell any surpluses already. This school argued that this formulation of the starting point would put unreasonable pressure on the system.

5.3.5 Flexibility with the Emissions Trading System

The European Council had also suggested introducing flexibility related to the EU ETS. This would be available only for higher income Member States and specifically those that foresaw difficulty in achieving their targets domestically.

Such Member States would be allowed to transfer a limited amount of allowances from the EU ETS, which they would normally auction, into the "Effort Sharing" and swap for allocations for the non-ETS sectors. As such, the Member States concerned would see their auctioning revenue decrease. The total amount of this flexibility is limited to 100 million allowances[20] over the period of ten years, which equates to the equivalent of less than 0.5% of the expected Annual Emissions Allocations in the non-ETS sectors over the period from 2021–2030.

The distribution of access to this flexibility provision within the group of higher income Member States followed the same reasoning as the adjustments of targets within this group. Ireland and Luxembourg, countries with the biggest projected gap with their targets are allowed the highest extent of access to this flexibility, equivalent to 4% of 2005 emissions *per annum*.[21] The UK, Germany and France have no gap in the projections, or a limited one, and get no access to this flexibility. All other higher income Member States (Netherlands, Belgium, Austria, Denmark, Finland, Sweden) have access equivalent to 2% of 2005 emissions *per annum*.

Malta is the only lower income Member State that can use this flexibility, at a rate of 2% of 2005 emissions *per annum*. The reason is that Malta is the Member State with the lowest *per capita* emissions in the non-ETS sectors, and this would result in emissions of below two tonnes per person in 2030. Together with the fact that it is the most densely populated Member State, Malta sees its mitigation target as potentially challenging. Allowing access to this one-off mechanism takes account of this specific circumstance.

5.3.6 Flexibility to land use change and forestry sectors

The second innovative flexibility with respect to the non-EU ETS sectors is related to "Land Use, Land Use Change and Forestry" (LULUCF). This establishes a link between the non-ETS sectors and the specific sectors regulated under the Regulation on the inclusion of greenhouse gas emissions and removals from land use, land use change and forestry in the 2030 climate and energy framework[22] (see also Chapter 8). This LULUCF Regulation foresees that Member States should ensure the absorptions and removals of emissions in the LULUCF sectors are not deteriorating compared to unchanged policy. Overall, it is expected that these sectors will actually absorb carbon, reducing atmospheric concentrations. If Member States perform better than expected this gives rise to the generation of LULUCF "credits," for instance due to the planting of new forests (afforestation) or due to the adaption of agriculture practices that improve the carbon retention of soil.

Under the Kyoto Protocol, and notably during the period from 2008–2012, Member States were allowed to fully use any such credits to compensate for emissions in other economic sectors. The LULUCF accounting rules that defined when credits could be generated were often seen as too lax, resulting in significant amounts of credits generated. Subsequently, under EU legislation for the period from 2012–2020, these credits could not be used to achieve targets in the EU ETS or non-ETS sectors. Several Member States saw this as an undue limitation of their ability to fulfil their climate change commitments and as a discouragement for additional action in the land use sector. Another group of Member States opposed the use of such credits in other sectors, fearing it would undermine the real reduction of emissions due to their temporary nature and thereby delay the necessary transition towards a low-carbon economy.

The legislation for 2020–2030 retained flexibility while addressing both concerns. First, the LULUCF accounting rules have built further upon those defined under the Kyoto Protocol but have improved them to avoid undue crediting. Second, while flexibility was allowed between the LULUCF sector and notably the non-ETS sector, a limitation was put on the quantity of credit allowed to be used for compliance purposes in the non-ETS sectors. This was set at 280 million tonnes for the EU over the ten-year period from 2021–2030, or on average 28 million per year. This will ensure that there is still need for strong reductions in the non-ETS sectors.

On the other hand, those Member States with a large agriculture sector have relatively greater access to these LULUCF credits. This recognises that agriculture is expected to reduce its non-CO_2 emissions less than other economic sectors by 2030 but, on other hand, it has more potential to take action to increase absorptions or reduce emissions in the LULUCF sectors. In effect, by this limited link, the non ETS sectors give further incentives to take action in the land sector, for instance to enhance the "sink" functions linked to agriculture.

All Member States have access to this flexibility, but they are grouped into three categories. Member States with a historic share of agriculture emissions in the non-ETS sectors of more than 25% have potential access to LULUCF credits equivalent to 15% of historic agriculture emissions. Only four countries qualify for this degree of access: Ireland, Lithuania, Denmark and Latvia. Member States with the lowest share of agriculture emissions in the non-ETS sectors only have potential access to the equivalent of 3.75% of the agriculture emissions; and the middle category to 7.5% of agriculture emissions. For a detailed overview of this distribution, see Table 5.2.

If used to its maximum extent, this flexibility represents the equivalent of 1% of the annual emissions of the non-ETS sectors in 2005. Furthermore, these LULUCF credits cannot be traded between Member States in the fulfilment of their non-ETS sector's obligations and can only be taken into account if a Member State would otherwise not be in compliance. These

TABLE 5.2 Distribution per Member State of maximum allowed LULUCF credits for potential use to comply with non-ETS targets

	Share of agriculture non-CO_2 emissions in non- ETS sectors 2008– 2012*	Average annual limit of LULUCF credits for compliance in the Non ETS as a % of annual 2008–2012 agriculture emissions	Estimate on maximum limit of LULUCF credits that can be used in the non-ETS over the period from 2021–2030 in million tonnes**	Average annual limit of LULUCF credits as a % of annual 2005 non-ETS emissions
EU	**16%**	**6%**	**−280**	**1.0%**
IE	40%	15%	−26.8	5.6%
LT	28%	15%	−6.5	5.0%
DK	27%	15%	−14.6	4.0%
LV	25%	15%	−3.1	3.8%
RO	24%	7.5%	−13.2	1.7%
BG	21%	7.5%	−4.1	1.5%
FR	20%	7.5%	−58.2	1.5%
EE	20%	7.5%	−0.9	1.7%
FI	18%	7.5%	−4.5	1.3%
ES	18%	7.5%	−29.1	1.3%
SE	16%	7.5%	−4.9	1.1%
CY	16%	7.5%	−0.6	1.3%
EL	16%	7.5%	−6.7	1.1%
PT	16%	7.5%	−5.2	1.0%
NL	15%	7.5%	−13.4	1.1%
SI	15%	7.5%	−1.3	1.1%
PL	15%	7.5%	−21.7	1.2%
HU	14%	3.75%	−2.1	0.5%
SK	14%	3.75%	−1.2	0.5%
UK	13%	3.75%	−17.8	0.4%
HR	13%	3.75%	−0.9	0.5%
AT	13%	3.75%	−2.5	0.4%
BE	13%	3.75%	−3.8	0.5%
DE	13%	3.75%	−22.3	0.5%
CZ	11%	3.75%	−2.6	0.4%
IT	10%	3.75%	−11.5	0.3%
MT	8%	3.75%	−0.03	0.3%
LU	7%	3.75%	−0.25	0.2%

* Rounded to the nearest percentage
** Calibrated to match 280 million tonnes

conditions further limit the extent to which LULUCF credits may be used, thereby reinforcing the safeguards of ensuring sufficient action is taken in the non-ETS sectors to reduce emissions.

5.3.7 Flexibility linked to earlier overachievement

One final additional flexibility provision was introduced to recognise early action in limiting emissions for Member States with income levels in 2013 below the EU average. It applies only to Member States that overachieve their targets in the period from 2013–2020, often generating considerable surpluses. The Regulation does not allow "carry over" of such surpluses to ensure the overall environmental integrity of the scheme. Instead in order to recognise these efforts, a limited "safety reserve" was created of a maximum of 105 million tonnes of CO_2-equivalent for the whole period from 2021–2030 to be distributed to those lower income Member States that do not achieve their 2026–2030 targets, proportional to their overachievement in the period from 2021–2030.

Instead, "safety reserve" can, however, only be used if the non-ETS sectors across the EU as a whole meet the EU's target for 2030. As this will be known only at the end of the period, it would be hazardous for any Member State to rely too much upon using previous overachievements (prior to 2020) in the period from 2021–2030.

5.3.8 The 2030 targets as adopted

The combined flexibilities described in this chapter reduce the extent to which certain Member States would need to depend on transfers from other Member States. Each Member State knows its 2030 target as well as the maximum amount of EU ETS and LULUCF credit flexibilities it is allowed to access. Figure 5.9 summarises the finally allocated 2030 targets and the access each Member States has to the LULUCF and EU ETS flexibility (expressed as an annual percentage of annual emissions *per annum*).

Conclusion: Differentiation of emissions reduction targets based on GDP *per capita* continues through to 2030, thereby considerably enhancing the convergence of emissions per person between the EU's Member States. Additional flexibility is introduced from 2021 without compromising the delivery, the overall fairness or as the cost-effectiveness of the EU's domestic target.

ESR targets and maximum one-off ETS/non-ETS and land use flexibilities

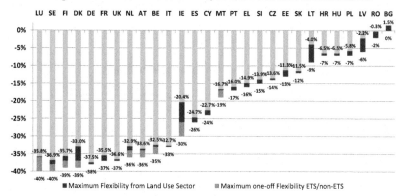

FIGURE 5.9 Member States' 2030 reduction targets in the non–ETS sectors as well as maximum amount of EU ETS flexiblity and LULUCF credit flexibility *per* Member State

Source: European Commission: https://ec.europa.eu/clima/policies/effort/proposal_en

Conclusion

, A comprehensive climate strategy for the EU as a whole required the incorporation of fairness and the accommodation of distributive impacts between the Member States. This has been a complex task, both politically and technically, for which the EU has been a pioneer, as no other countries have such legally binding targets in place. Fairness is itself not simple to ensure, yet it has been introduced through the differentiation of obligations, as well as with respect to the flexibilities allowed. As a result, 28 sovereign Member States could unanimously agree on one common climate policy, despite their different levels of economic development, different industrial strategies and different energy systems with varying degrees of dependence on fossil fuels.

The "Burden Sharing" of the Kyoto Protocol's first commitment period (2008–2012) was decided before the EU ratified the Protocol in 2002. For the 2020 targets it was the Effort Sharing Decision that continued this process, although materially different in that the scope excluded those sectors covered by the EU ETS. The latest phase of this fairness exercise is the Effort Sharing Regulation from 2021 to 2030, adding flexibilities driven by the need to take account of specific circumstances of Member States, as well as recognition of efforts in the land use sectors, while limiting the risks to environmental integrity.

At the same time, the overall level of ambition was increased substantially, in line with the contribution under the Paris Agreement, but still underpinned by solid considerations of cost-efficiency. The sharing of obligations was informed and refined on the basis of economic modelling and factual analysis. One important prerequisite was the availability of clear, straightforward factual information that the EU had already developed as part of its obligations under the Kyoto Protocol.

The bottom-up nature of the Paris Agreement allows for significant differentiation, but significantly more ambition will be needed in the future. The EU's example of differentiation could be informative for other big nations with a federal structure to develop internal policies in a similar manner. Equally, groups of nations could act together and differentiate their efforts, but this requires political trust and a common governance system to provide solid and comparable data, elements of which are provided for under the "rulebook" of the Paris Agreement as agreed at COP24 in Katowice.

Notes

1 "Decision No 406/2009/EC of the European Parliament and of the Council of 23 April 2009 on the effort of Member States to reduce their greenhouse gas emissions to meet the Community's greenhouse gas emission reduction commitments up to 2020", OJ L 140, 5.6.2009, pp. 136–148. http://eur-lex.europa.eu/LexUriServ/LexUriServ.do?uri=OJ:L:2009:140:0136:0148:EN:PDF.

2 "Regulation (EU) 2018/842 of the European Parliament and of the Council of 30 May 2018 on binding annual greenhouse gas emission reductions by Member States from 2021 to 2030 contributing to climate action to meet commitments under the Paris Agreement and amending Regulation (EU) No 525/2013", OJ L 156, 19.6.2018, pp. 26–42. See: https://eur-lex.europa.eu/legal-content/EN/TXT/PDF/?uri=CELEX:32018R0842&from=EN.

3 Council Decision 2002/358/EC of 25.4.2002.

4 These limits were specified in Article 5 of Decision No. 406/2009/EC of the European Parliament and of the Council of 23 April 2009 on the effort of Member States to reduce their greenhouse gas emissions to meet the Community's greenhouse gas emission reduction commitments up to 2020. Official Journal L 140 of 5.6.2009, pp. 136–148.

5 See footnote 1.

6 The selection of 2005 as a base year was because it is the first year data is available for the ETS installations and thus the first year for which the split can be made between ETS and non ETS GHG emissions in the EU at Member State level.

7 GDP *per capita* measured at market prices.

8 European Commission Staff Working Document SWD (2016)247 final of 20.7.2016.

9 SEC(2008)85/3, Impact Assessment accompanying the Package of Implementation measures for the EU's objectives on climate change and renewable energy for 2020.

10 European Commission website on harmonised standards detailing the relevant Ecodesign and energy labelling framework legislation and implementing measures: https://ec.europa.eu/growth/single-market/european-standards/harmonised-standards/ecodesign_en.

11 https://ec.europa.eu/growth/single-market/european-standards/harmonised-standards/ecodesign_en.

12 Model-based Analysis of the 2008 EU Policy Package on Climate Change and Renewables by P. Capros, L. Mantzos, V. Papandreou, N. Tasios basis on the PRIMES model - E3MLab/NTUA, (June 2008) for the European Commission (DG Environment): https://ec.europa.eu/clima/sites/clima/files/strategies/2020/docs/analysis_en.pdf.

13 Report by the European Commission: "Two Years After Paris – Progress Towards Meeting the EU's Climate Commitments". COM(2017)646 final of 7.11.2017, p. 5.

14 Commission Staff Working Document SWD (2018)453 final of 26.10.2018 (PART 6/6): Technical Information Accompanying the document Report from the European Commission to the European Parliament and the Council *"EU and the Paris Climate Agreement: Taking Stock of Progress at Katowice COP"*(COM(2018)716final), Section 7, Table 4. (See: https://eur-lex.europa.eu/resource.html?uri=cellar:e20ad4bb-d933-11e8-afb3-01aa75ed71a1.0001.02/DOC_6&format=PDF).

15 European Commission Staff Working Document SWD (2016)247 final of 20.07.2016.

16 "Regulation (EU) 2018/842 of the European Parliament and of the Council of 30 May 2018 on binding annual greenhouse gas emission reductions by Member States from 2021 to 2030 contributing to climate action to meet commitments under the Paris Agreement and amending Regulation (EU) No 525/2013", OJ L 156, 19.6.2018, pp. 26–42. See: https://eur-lex.europa.eu/legal-content/EN/TXT/PDF/?uri=CELEX:32018R0842&from=EN.

17 An AEA represents one tonne of CO_2-equivalent.

18 Although the European Commission had proposed a trajectory starting in the year 2020, the finally adopted compromise by the European Parliament and Council put the starting point at "five-twelfths of the distance from 2019 to 2020 or in 2020, whichever results in a lower allocation for that Member State."

19 Commission Impact Assessment on the proposed Effort Sharing Regulation (SWD (2016)247 final of 20.7.2016).

20 An allowance represents one tonne of CO_2-equivalent.

21 So the flexibility is equivalent to ten times 4% of its 2005 emissions in the non-ETS sectors.

22 Regulation (EU) 2018/841 of the European Parliament and of the Council of 30 May 2018 on the inclusion of greenhouse gas emissions and removals from land use, land use change and forestry in the 2030 climate and energy framework and amending Regulation (EU) No 525/2013 and Decision No 529/2013/EU. https://eur-lex.europa.eu/legal-content/EN/TXT/PDF/?uri=CELEX:32018R0841&from=EN.

6

ENERGY-RELATED POLICIES AND INTEGRATED GOVERNANCE

Artur Runge-Metzger, Stefaan Vergote and Peter Vis

Introduction

Although energy policy has been at the core of the European project since the establishment of the European Coal and Steel Community (ECSC) and European Atomic Energy Community (EURATOM), both in the late 1950s, it was not until the amendments to the Treaty agreed in Lisbon in 2009 that a provision specifically on energy was included. Article 194 of the Treaty on the Functioning of the European Union links the Union's energy policy with the "need to preserve and improve the environment."

The Articles on the environment in the Treaty on the Functioning of the European Union had been introduced much earlier, by the Single European Act in 1986. Although most measures relating to the environment are decided by the ordinary legislative procedure and qualified majority in the Council, Article 192, relating to the environment, also referred to decisions being taken unanimously on "measures significantly affecting a Member State's choice between different energy sources and the general structure of its energy supply." Unlike environmental measures, therefore, decisions related to the energy mix were singled out for decision by unanimity.

From the start, climate change and energy policies were inextricably linked because of the importance of the use of fossil fuels. This coincided with the common concern as regards the EU's dependence on imported energy. In 2015, 88% of oil, 69% of gas and 64% of coal was imported, representing net fossil fuel dependency of 79%. By 2030, imports could even increase to 94%

for oil and 83% for gas. An integrated climate and energy policy could help reduce the EU's over reliance on fossil fuels.

The five-year mandate of the European Commission that started in 2015 included amongst its ten political priorities, "a resilient Energy Union with a forward-looking climate change policy." The aim of this priority was to strengthen the coherence between the energy and climate files, as well as to "mainstream" climate change into most other EU policies. This chapter focuses on how the policy toolbox has evolved towards making energy more secure, affordable and sustainable.

6.1 Renewable energy

6.1.1 A binding EU-wide target

Along with its target for greenhouse gas emissions, the European Union set itself a target of 20% renewables in the energy mix by 2020. In 2007, the European Council also agreed that this 20% renewable energy target should be translated into binding renewable energy targets for each of the Member States, "taking account of different national starting points and potentials, including the existing level of renewable energies and energy mix."[1]

The Commission's initial approach to address fairness was to estimate the technological potential of each Member State to define such national renewable energy targets. However, this approach suffered from a lack of transparent data that the Member States could easily agree with. Therefore, fairness was implemented in another way. First, the "gap" was calculated between where the EU was at the time (8.7% renewable share in 2005) and its target of 20% in 2020. Half of this gap of 11.3% would be shared equally across all Member States (a flat rate of 5.75% was used), and the other half was shared amongst Member States on a GDP *per capita* basis. This implemented "fairness" insofar as every Member State had an element of the same flat-rate increase, seen by some as fair, combined with each country having a GDP *per capita* component that reflected its relative wealth, seen by others as fair. In this approach, richer Member States had to do more.

Mindful that these renewable targets had not been determined in such a way as to ensure their cost-efficiency, the Renewable Energy Directive of 2009 created cooperation mechanisms, whereby Member States would be able to reallocate overachievement by one in favour of an under-achievement by another – implicitly in exchange for payment. It was well understood that the marginal cost of achieving the last 1% towards the 20% goal would be considerably higher than the 1% increase from, say, 9% to 10%. To date, these cooperation mechanisms have not been frequently used, but they remain available for the fulfilment of national targets through to 2020 and beyond. Some Member States are indeed overachieving and others under-achieving their 2020 targets, so the use of cooperation mechanisms is likely to increase (see Figure 6.1).

The policy of translating the overall EU target into binding contributions for each Member State worked well. In 2017, renewable energy had a 17.5% share of final energy consumption, which compares well with its target of 20% by 2020. Thanks to the significant over-delivery in a few Member States, the 20% target is likely to be achieved. In 2014, however, the European Council decided on an EU target for renewable energy of 27% for 2030 and dropped the need for a binding renewable energy target for each Member State. As time progressed, the legislators decided in 2018 to increase the EU-wide target for renewable energy in 2030 to 32%, with a possible upward review by 2023.[2]

The explanation for this change of approach is twofold. First, Member States were still free to set such binding national targets for themselves if they wished to. For 2020, the EU fixed an overall target for renewable energy and greenhouse gas reductions and gave each Member State individual legally binding targets in order to reach the overall EU renewable energy and climate targets. In the 2030 perspective, Member States have been given greater freedom in how they meet their climate targets, wishing to reach them at

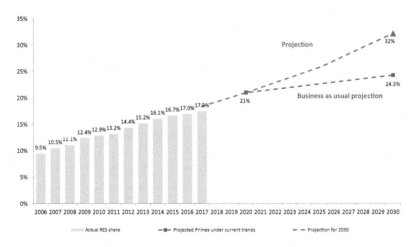

FIGURE 6.1   The development of renewable energy in the EU, 2005–2030

Source: Eurostat and PRIMES model BAU projections as at 2016

the lowest cost possible. The political reality was also that there was a lack of sufficient support for national binding renewable targets. The delivery of the 2020 renewable energy targets by Member States had required significant financial incentives, agreed in the context of the EU budget negotiations, which led to debate and sometimes controversy. Today, however, this situation has changed significantly insofar as the additional cost of renewable energy has fallen substantially, as illustrated in Figure 6.2. This cost reduction happened in large part through the significant expansion of the market, not least as a result of the existence of binding national targets for renewable energy.

The other reason for the change in policy with regard to binding renewable energy targets was that the renewables were increasingly affecting power generation markets and networks. The emergence of renewable electricity challenged the business model of fossil and nuclear generators and Member States needed to take a range of steps to integrate renewables, while maintaining the continuity of electricity supply. By 2014, renewable energy had become a significant player instead of an isolated future-oriented niche. The internal market was playing an increasing role in electricity markets, but not in renewable energy support schemes, that were designed within each Member State and tailored to national circumstances. The policy question shifted

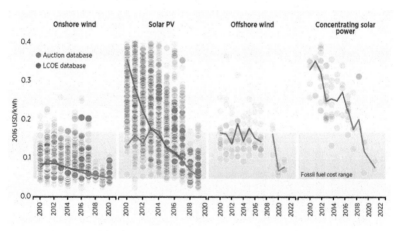

FIGURE 6.2 Global levelised cost of electricity and auction price trends from renewable energy sources, 2010–2020

Source: IRENA Renewable Cost Database and Auctions Database

from questions about optimal support structures for renewable energy to questions relating to the organisation of the energy markets. The changed market situation led to the conclusion that Member States would anyway need to invest heavily in low-carbon technologies in order to meet their greenhouse gas reduction targets for 2030, and that binding renewable targets for each Member State were no longer a necessity in this context.

6.1.2 Biofuels

Biofuels have been a major part of the discussion on renewables as they were considered the lowest cost, large-scale "solution" in the short term to the transport sector's ever-increasing emissions of greenhouse gases. The initial assumption was that the carbon in the plants or trees had been absorbed from the atmosphere, and burning the fuel was only putting carbon back into the atmosphere where it had come from. This is why the IPCC Guidelines report CO_2 emissions from the combustion of biomass as zero *in the energy sector*. At the outset, therefore, biofuels were presumed to be good. It was rather quickly realised that there are many more aspects to consider beyond the energy sector, such as that cultivation may entail energy intensive fertilisers, or because land is scarce, may cause displacement effects on land use in Europe or elsewhere in the world.

The European Union first introduced a blending target in 2003 with the aim to reach a 5.75% share of renewable energy (essentially biofuels) in the EU's transport sector by 2010.[3] As part of the EU's 2020 objectives, the target was increased to 10% and it was made mandatory for each Member State.[4] For the first time, sustainability criteria were also introduced to respond to questions on the net benefits in terms of reducing greenhouse gas emissions. These criteria required biofuels not to be grown on land with a high carbon stock or high biodiversity, such as primary forestland. In addition, greenhouse gas savings compared to fossil fuel were required of at least 35% from 2009 and 50% from 2017, taking into account emissions arising during their life cycle, for example cultivation of the raw materials, or "feedstock," processing and transport. In 2015, new legislation was introduced to take account of emissions resulting from Indirect Land Use Change (ILUC) effects.[5] The use of food- and feed-based biofuels was capped at 7%.[6] In addition, biofuel production from new installations were required to ensure a greenhouse gas saving of at least 60% to comply with the sustainability requirements.

The legislation agreed in 2018 incorporates a 14% sub-target for renewable energy in transport but stipulates that not more than 7% can be realised through so-called first-generation biofuels produced by the agricultural sector, such as from rapeseed.[7] In addition, a greenhouse gas saving of 65% is required of sustainable biofuels from 2021. There are no longer any binding targets specifically on Member States for renewable energy in transport, but the 14% target is a requirement Member States must impose on fuel suppliers. The emphasis is explicitly to increase the production of advanced biofuels from the recycling of waste material such as cooking oil and animal fat, or from cellulosic wood material. Furthermore, new technological breakthroughs can be expected through the use of biofuels that are not land-based, such as algae. Several "multipliers" are foreseen as an additional incentive to the use of the most costly advanced biofuels, as well as for renewable electricity used by the rail and road sectors.[8] Finally, a process has been agreed for the establishment of a certification process that is intended certify food-crop biofuels that are deemed to have a low "indirect land use change," or displacement, effect. Food-crop biofuels that are deemed to have a "high indirect land-use change-risk" will be completely phased out by 2030. Palm oil, for example, is claimed to be such a high-risk feedstock. This certification process has begun and is being completed. Overall, the legislative framework seeks to ensure that biofuel use unequivocally contributes to emission reductions.

EU biofuels policies have gradually led to more serious reflection on emissions beyond the traditional energy sector, including accounting and mitigating emissions in forestry, agriculture and for land use change in general. It must be admitted, however, that initially the EU rather rushed into biofuels policy without full consideration of all possible impacts, while the benefits for climate change were exaggerated at the outset. Even now, with all the environmental safeguards in place in Europe, some still dispute the added value of biofuels, arguing that they are energy intensive to produce, used in inefficient internal combustion engines and throw a lifeline to the fossil fuel-based products, with which biofuels are blended. From an air quality perspective, biofuels are of little benefit. From a climate change perspective, their production can be worse than fossil fuels if their production encourages deforestation through land-use displacement effects, not only within the EU but also, more importantly, in Asia, Latin America or Africa, as food and feed commodity markets are truly global. In addition, biofuels produced from food and feed feedstock could increase food prices.

It is understandable that, in view of these doubts, legislators have increasingly concentrated promotion initiatives and provisions on fuels where there is greater consensus on their being beneficial for the environment. Few parliamentarians and few members of the public have the opportunity to become experts in this field; it is understandable that in light of such controversy, as has been seen in recent years, there has been a relocation of support for biofuels onto a concentration of promotional efforts on what are clearly – and consensually – considered to be beneficial for the environment. In the end, it is now widely felt that subsidies, incentives and research should rather be used to promote alternative powertrain technologies, in particular electric technologies, for which sustainable renewable energy is increasingly available. Production of sustainable liquid fuels, including synthetic fuels from renewable energy, is possible and is still being developed, but their production will be limited in scale. Such sustainable liquid fuels should rather be reserved for uses where technological alternatives are not yet available, such as in aviation.

6.1.3 Biomass

A particular aspect of renewable energy policy is the promotion of biomass as a fuel. In analogy with biofuels, the 2018 Renewable Energy Directive introduces sustainability criteria for the first time. This is necessary as the use of biomass gets a double incentive: under the renewable energy target as well as under the EU ETS where its use is deemed "emissions free." However, probably the most important new element as far as biomass is concerned are the new accounting rules set under the Land Use, Land Use Change and Forestry (LULUCF) Regulation (see Chapter 8). In particular, sustainable forest management and afforestation make a positive contribution to climate change if done properly.

There are doubts, however, relating to the sustainability of biomass for energy, especially when land use, agriculture and forestry emissions are taken into account.[9] Clearly, there is a high risk of problems if biomass is sourced from a country that is not committed to account for its forestry emissions under the Paris Agreement's greenhouse gas accounting rules. Therefore, the Renewable Energy Directive foresees that the import of biomass material must come from countries with LULUCF accounting in place. In the future, one can also expect satellite monitoring to play an increasing role in the overall accounting of this most important sector, not least at the international level.

The future use of biomass needs to be considered in conjunction with newly emerging carbon capture and storage technologies as potential means of absorbing and storing carbon dioxide that would otherwise persist in the atmosphere. If Europe and other countries want to meet the temperature goals of the Paris Agreement for the second half of the century, these sequestration and capture techniques and technologies must be nurtured from a much earlier stage to reach maturity in time. There are scenarios where carbon can be absorbed out of the atmosphere by biomass: after the use of biomass as a combustion fuel, the carbon is separated or captured through the appropriate technology, and finally stored underground. This so-called "BECCS" technology, which stands for Bio-Energy with Carbon Capture and Storage, has the potential to produce "negative" carbon dioxide emissions in combination with the new LULUCF management and accounting techniques. The IPCC refers to this type of technology in its recent reports, such as its 1.5°C report of 2018, without giving it an endorsement. However, it does have the theoretical potential for the bio-economy to make a worthwhile contribution to the stabilisation of climate change globally.

Conclusion: The EU's renewable energy target of 20% by 2020 is being delivered and an increase in ambition up to 32% by 2030 has been agreed. The expansion of the renewable energy market has led to significant cost reductions. From now on electricity market dynamics, including the carbon price, will be the driving force behind further investments in renewable energy. The use of biofuels and biomass are subject to strict sustainability criteria.

6.2 Electricity market integration and the Market Design Initiative

The substantial increase in the share of electricity generated from renewable energy sources together with the gradual opening of the electricity market required a fundamental review of the rules managing the EU energy market. In November 2016, the European Commission therefore released a set of proposals, referred to as the Market Design Initiative,[10] which is another major opportunity to facilitate the ongoing sustainable energy transition.

6.2.1 The challenge of integrating renewable energy

Both wind and solar energy are variable sources of energy. Wind availability tends to be uncorrelated with demand and is prone to unpredictable fluctuations in its intensity, requiring intervention by the electricity system operators. While solar energy is more predictable, the sunrise and sunset effects require significant adjustment across the network, other types of generation having to come on line to replace it in order to satisfy demand. These adjustments have had impacts on conventional generation assets, most of which were built to operate at relatively constant level throughout the day, without fast ramping capabilities. There can also be relatively long periods with very little or a great deal of wind and sun. In such cases, other sources of energy are needed to replace missing generation, or conversely, during periods when renewable energy is abundant, conventional assets may be standing idle.

In addition, the zero marginal cost of variable renewable energy has led to very low, sometimes even negative, wholesale market prices. In combination with the subsidies for renewable energy, the wholesale market revenues of power companies are significantly reduced. In these circumstances, the economic viability of new investments is adversely affected, putting at risk the resilience of the electricity system as a whole. To maintain investment in capacity needed to meet peak demand at times when renewable energy is limited due to weather conditions, Member States resorted to so-called capacity mechanisms that have served to fragment European electricity markets.

One of the "remedies" introduced in 2009 to enable renewable electricity to compete in recent years has been "priority dispatch," where Transmissions System Operators had an obligation to first take the renewable energy that was available, if necessary at the expense of conventional fuels. However, due to these deliberate market interventions, electricity markets and investments were distorted. Consequently, in the recent reform of electricity markets, the notion of "priority dispatch" for renewable energy was discontinued. Instead, it was decided that all wholesale market participants should face the same responsibilities in terms of grid balancing in order to ensure the better integration of renewable energy supply, demand response and storage solutions, as well as reserve capacity waiting on stand-by in the case of unplanned events. The emphasis is now on regional cooperation, the more efficient use of interconnectors and the procurement of reserve needs. So rather than concentrating solely on increasing the share of renewable energy, policy now takes a wider view in order to ensure that the electricity supply system as a whole remains strong and resilient.

6.2.2 The combined effects of electricity market reform and carbon pricing

A well-functioning electricity market, in combination with a well-functioning EU ETS, can give sufficient price signals for the long-term investments consistent with Europe's decarbonisation goals.[11] As carbon prices have returned to their pre-2010 levels and technology costs fall, there will be a gradual shift from conventional sources to renewable sources as well as a shift from coal to gas.

This combination of carbon and electricity markets becomes the driver for the energy transition in the power generation sector. The auction prices of recent renewable energy tenders for photovoltaics, onshore and even off-shore wind suggest that this transition is achievable, especially if carbon prices continue to recover. In order for consumers to be incentivised to save energy, the Commission proposed to phase out retail price regulation, still present in a number of Member States. Social tariffs will still be allowed, subject to certain requirements and price regulation in emergencies is still permitted. Energy poverty will be measured at the national level and its root causes addressed, such as through targeted energy efficiency measures. Fairness is not only about managing the cost of energy for those who cannot afford it, but it is also about expecting those who can afford energy to pay a price that takes externalities – such as its effects on the environment – into account.

Finally, subject to strict conditions, it is still possible for Member States to introduce capacity mechanisms to address security of energy supply, subject to all other market design improvements having been implemented. However, a maximum threshold of 550 grams CO_2/KWh that a generation plant can emit is set in order to be able to receive payments from national capacity mechanisms. This threshold excludes conventional coal-fired generation and for this reason has been heavily criticised by certain Member States. However, the main aim is to avoid support being given to investments in highly emitting generation assets that are inconsistent with the EU's long-term decarbonisation policy. This could otherwise very likely lead to having a number of costly "stranded-assets" in the power system because of early retirement due to their incompatibility with overriding goals. To date, at least eleven Member States have introduced and had a variety of capacity mechanisms approved by the European Commission. These are generally technology neutral[12] and mostly ensure gas and coal power stations are available in the event of shortages of electricity supply.

6.2.3 Strengthened role for consumers

Recognising the importance of energy choices made by citizens, the opportunities for consumer choice and engagement with energy retail markets has been increased. Information provided to consumers, such as on electricity bills, will be improved and the possibility offered to participate in energy markets directly or through companies that represent them, such as "aggregators." Moving from one energy supplier to another will be facilitated, creating more price competition at the retail level. Every customer can request a smart meter to better manage energy consumption and benefit from a dynamic price contract if they want one. The installation of smart meters is also important for promoting the participation of consumers as generators of electricity for their own consumption, or for selling, storing or offering to change consumption patterns as part of a demand-response programme, receiving remuneration either directly or through aggregators. All of these measures will contribute to energy security and enhance energy efficiency, while potentially enabling consumers to save money.

Conclusion: In combination with a reinforced EU ETS, the energy market reforms encourage clean-energy investments, fuel switching and grid stability. Distortions needed to be removed from electricity markets so that renewable energy, energy efficiency measures, demand response, as well as energy storage are all incentivised.

6.3 Energy efficiency

6.3.1 Energy dependence, the import bill and barriers to energy efficiency

Concerns persist around the EU's dependence on imported energy. Best, of course, would be not to need so much energy: it is often said that the most efficient and cleanest energy is energy that is not consumed. Energy efficiency results in less energy being used and lower energy bills for consumers. However, despite significant technological progress in the field of energy efficiency and its societal benefits, the level of investment in energy efficiency in Europe is still below its economic and technological potential.[13]

Energy efficient investments with a payback time of four or five years are often not undertaken in both the private and public sectors. Market and behavioural barriers such as imperfect information, split incentives or the distrust of the energy-efficiency business model with its upfront costs hinder consumer uptake of energy efficiency measures, notwithstanding lower energy bills and other societal benefits. Governments and public sector actors may be reluctant to undertake energy efficiency programmes due to the pressing need to consolidate public finances. Individual consumers may lack access to attractive financing terms, they may be reluctant because of the possibility of moving house or they may incur the inconvenience of building work being done. There are still many hurdles to jump between knowing what can be done and actually doing it: as in so many other areas of life, the choice between short-term convenience and long-term reward does not always favour the longer-term gains.

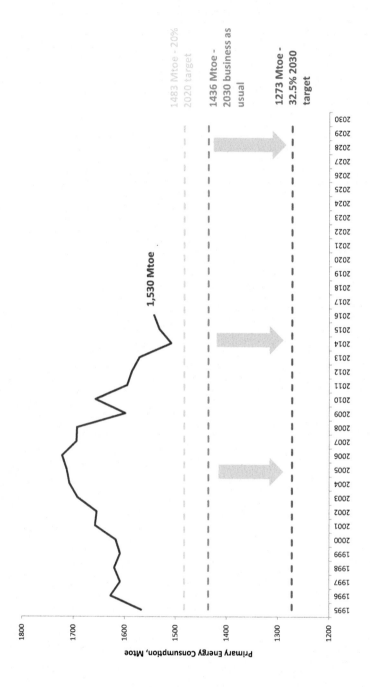

FIGURE 6.3 Ambition level of EU's 2030 energy efficiency target of 32.5% compared to primary energy consumption and "business-as-usual" forecasts

Source: Eurostat and PRIMES model BAU projections as at 2016

There are, therefore, solid arguments for government intervention and regulation to make things happen. The European Union has advocated the "Energy efficiency first" principle and agreed for 2030 an energy efficiency improvement target of 32.5% at the EU-level. This target is defined as a 32.5% reduction of gross primary energy consumption compared to the 1887 million tonnes of oil equivalent (Mtoe) level that was expected for 2030 under a "business-as-usual" projection made in 2007.[14] The more recent projection done in 2016 estimates the primary energy consumption of the EU at much lower level of 1436 Mtoe. Hence, a 32.5% energy efficiency improvement seems feasible and maintains the consistency with the EU's climate and renewable energy objectives.

6.3.2 The EU's approach to energy efficiency targets and policies

The EU overall target on energy efficiency has not been translated into legally binding targets for Member States but will be realised through the newly established governance system. There were several reasons for this. First, Member States were already obliged to limit their greenhouse gas emissions and improving energy efficiency would help to meet this target. In addition, from a macroeconomic standpoint, improving energy efficiency is very strong candidate measure. What is needed above all, however, are common policies rather than targets, especially as the appliances and vehicles that used energy were so often manufactured in (or exported to) other Member States, and individual governments were unable to unilaterally regulate product standards in the context of Europe's internal market.

The EU has made substantial progress towards its energy efficiency objectives, but it is not on track to meet its energy efficiency target for 2020. While energy savings have helped offset the impact of increases in economic activity, in recent years they have not been enough to offset the increase in energy consumption. Although it was never specifically agreed whether the EU's 2020 energy efficiency target should be defined in final or primary energy consumption terms,[15] the European Commission measures progress against both definitions. Final energy consumption in the EU fell by 5.9%, from 1193 Million tonnes of oil equivalent (Mtoe) in 2005 to 1122 Mtoe in 2017. This is still 3.3% above the 2020 final energy consumption target of 1086 Mtoe. Primary energy consumption in the EU dropped by 9.2%, from 1720 Mtoe in 2005 to 1561 Mtoe in 2017. This is 5.3% above the 2020 target of 1483 Mtoe. Primary energy consumption decreased on average by 0.8% per year between 2005 and 2017, but it has been rising again since

2015. A year-by-year increase of 0.9% was recorded in 2017.[16] Further action is therefore clearly required at all levels of governance: European, national, regional and municipal.

At the European level, the approach has been to pursue the following four policies. First, to set performance standards for newly traded goods for both environmental and internal market reasons. Second, to set performance standards for new buildings for environmental reasons and create demand for better-performing building materials. Third, to provide information to consumers on the energy performance of all buildings, the labelling of new appliances and the fuel consumption/CO_2 performance of cars; and finally, to create obligations for energy distribution companies to save energy, for larger companies to undertake regular energy audits and for Member States to submit plans and develop national strategies.

This approach was implemented through the Ecodesign Directive,[17] the Energy Labelling Directive,[18] the Energy Performance of Buildings Directive[19] and the Energy Efficiency Directive.[20]

6.3.3 Regulating the energy use and labelling of products and devices

The "Ecodesign Directive" sets common performance standards, from an environmental perspective, of energy-consuming goods sold in the European Union. Many different categories of electrical and electronic equipment are covered, including heating equipment. The rationale for this legislation is, of course, to save energy and reduce emissions but also to avoid differences in national laws that would obstruct intra-EU trade. The Commission estimates that the Ecodesign Directive contributes around half of the energy savings target for 2020.[21]

The Ecodesign Working Plan for 2016–2019, now being deployed, includes a list of new product groups such as building automation and control systems, electric kettles, lifts, refrigerated containers, hand dryers, high-pressure cleaners or photovoltaic systems. Further new actions include setting the minimum energy efficiency requirements for air heating and cooling products and a measure to strengthen verification tolerances that relate to the compliance checks performed by Member States.[22]

Since 1992, there is also an Energy Labelling Directive. The energy label is a tool to assist consumers when purchasing household appliances, such as, washing machines and dishwashers. The eight categories "A" to "G" (more recently "A+++" to "D") show to what extent the product is economical and environmentally friendly. The Directive intends to offer critical information to consumers on the energy use of products, enabling a more informed choice, lower energy bills and ultimately fewer CO_2 emissions.

6.3.4 Addressing the energy efficiency of buildings across the EU

The Energy Performance of Buildings Directive is another important piece of legislation. Residential and commercial buildings are large users of energy and buildings account for some 40% of energy consumption in the EU. Under this Directive, Member States must establish and apply minimum energy performance requirements for new and existing buildings, ensure the certification of these energy performance requirements and ensure the regular inspection of boilers and air-conditioning systems in buildings. Moreover, the Directive requires Member States to ensure that by 2021 all new buildings will effectively become "nearly zero-energy buildings."

The 2018 revision of the Directive aims to increase building renovation rates and foster the delivery of smart building technologies. It puts stronger emphasis on Member States' renovation strategies aiming to decarbonise building stocks by 2050, including new provisions to promote building automation in conjunction with digital technologies. Importantly, it also requests measures to allow for the installation of charging points for electric vehicles, anticipating the need to decarbonise transport.

The proposed measures will generate a considerable investment effort amounting to some €80 to €120 billion in 2030. The Smart Finance for Smart Buildings Initiative[23] accompanies the Directive. It consists, for example, of support for the aggregation of dispersed small-scale investments and a de-risking pillar that aims to reduce perceived risks by investors.[24] The initiative builds on the previous Private Finance for Energy Efficiency Instrument, which is already operational in nine Member States, and is implemented in cooperation with the European Investment Bank.

6.3.5 The Energy Efficiency Directive

Finally, the Energy Efficiency Directive was reinforced in 2018[25] and sets an energy efficiency target for the EU of 32.5% in 2030, along with a potential upward revision by 2023. It includes a wide range of policy measures, including provisions relating to residential energy efficiency, smart meters, home energy management, energy audits in the commercial sector, retrofitting of public buildings, district heating and demand response.

According to Article 7 of the Directive, Member States have to introduce legislation that obliges energy distributors and suppliers to achieve savings of on average 1.5% a year of energy sales to final consumers from 2014 and 2030 or introduce alternative measures with the same effect. This is innovative, as it will make energy suppliers *de facto* suppliers of energy efficient services and products. This is also significant as nearly half of the energy savings from the Directive are expected to come from this Article alone.[26]

Conclusion: By 2030, energy efficiency will be improved by 32.5% in relation to a pre-determined baseline. Several measures aim to save energy but also to bring important benefits in terms of security of energy supply, improvement in the EU's trade balance and reduction of greenhouse gases.

6.4 Strengthened governance of the Energy Union

The 2014 European Council conclusions also call for a reliable and transparent governance system for the Energy Union to ensure that the EU meets its climate and energy policy targets. All levels of government, whether European, regional, national or local should contribute to this task, and the main tool to coordinate policies will be an integrated climate and energy plan to be prepared by each Member State.

The Governance of the Energy Union Regulation[27] builds upon and integrates the Monitoring Mechanism Regulation (MMR) for greenhouse gases and existing requirements for planning, reporting and monitoring in the climate and energy fields. It will reduce the administrative burden for Member States and the EU to comply with the reporting obligations under the UNFCCC and the Paris Agreement. The Regulation also sets an outlook for longer-term climate and energy action until 2050.

6.4.1 The EU's Monitoring Mechanism Regulation

The EU's Monitoring Mechanism Regulation (MMR)[28] determines the EU's internal reporting rules on greenhouse gas emissions. It is based on internationally agreed obligations under the UNFCCC and the Kyoto

Protocol. The Monitoring Mechanism Regulation was the first policy decision to be adopted at EU level on climate action in the early 1990s and has been considerably developed since then. It will continue to apply until 2021 when it will be incorporated into the Governance Regulation.

The Monitoring Mechanism Regulation enables the EU to have accurate annual information on greenhouse gas emissions and climate action. Member States are requested to report on past emissions from all economic sectors, projections of how emissions are expected to develop in the future, policies and measures to cut greenhouse gas emissions, climate adaptation measures, low-carbon development strategies, financial and technical support to developing countries, as well as Member States' use of revenues from the auctioning of EU Emissions Trading System allowances. By making this information widely and publicly available, it serves as a transparent basis for further research work and policy development.

6.4.2 Integrated climate and energy plans – the novelty of the Energy Union

Climate and energy policies are interlinked. Energy efficiency and renewable energy policies are key to promoting the achievement of greenhouse gas emissions reduction targets. They are of equal importance in the transition towards a more competitive, secure and sustainable energy system.

Integrated national energy and climate plans are at the core of the Energy Union Governance system. Member States will have to adopt such plans for the first time by December 2019 for the years 2021–30.[29] The plans and their reviews are to be synchronised with the Paris Agreement's five-year review cycles and associated global stock takes.

These integrated national energy and climate plans will lay out projections and objectives for the five dimensions of the Energy Union,[30] together with the policies and measures intended to achieve them. These plans will be comprehensive and should include transport, environmental, research and competitiveness aspects as well as removals through sinks. The draft plans will provide the EU and other Member States with an early indication on whether national efforts will be sufficiently ambitious to meet the Energy Union objectives, in particular the EU 2030 climate and energy targets. Templates have been agreed both to assist the Member States and to make the plans comparable between them.

Furthermore, the Commission may make recommendations to Member States. These recommendations are modelled on and complementary to those of the European Semester, which focuses on macroeconomic and structural reforms whereas the Governance Regulation addresses energy and climate specific policy issues. Recommendations made in the context of Article 194 of the Treaty on the Functioning of the European Union will be non-binding, but Article 9(3) of the Regulation stipulates that,

> Each Member State shall take due account of any recommendations from the Commission in its integrated national energy and climate plan. If the Member State concerned does not address a recommendation or a substantial part thereof, that Member State shall provide and make public its reasons.

Conclusion: Member States are required to prepare comprehensive and integrated energy and climate plans. These plans will include projections and provide the basis for the monitoring of progress towards meeting 2030 targets and the long-term goals of the Paris Agreement.

Conclusion

Climate and energy policies are inextricably linked, and over time, a more integrated approach has emerged across Europe. That is the best guarantee for a successful low-carbon energy transition, building less and less on fossil fuels and more on energy efficiency and renewable energy. This also helps to address the very high dependence of Europe on the import of coal, oil and gas.

Economic modelling has demonstrated that internalising environmental externalities is of primary importance. Equally, overcoming the barriers to improving energy efficiency and enabling dispersed renewable energy sources is critical to compete with large fossil fuel incumbents. Grid resilience in an electricity system with increasing amounts of variable renewable energy needs to be enhanced, as does the development of new technologies such as energy storage or rapidly spreading digitalisation across the energy sector.

Measures taken at the European level still leave plenty of scope for national and municipal measures through tax incentives (e.g., zero-energy houses being subject to lower property taxes) and subsidies (e.g., for better housing insulation), energy audits for private households, or the installation or upgrading of district heating systems. The concept of national integrated plans developed with regional and municipal authorities, as well as other stakeholders, is designed to help align policies and initiatives at all levels of governance.

Climate and energy policies have been made to work better together over time, such as with the introduction of the Market Stability Reserve into the EU Emissions Trading System, which improves its complementarity and compatibility with energy efficiency and renewable energy policies. The Market Design Initiative improves the overall system integrity, giving even small renewable energy producers the right to produce their own electricity and sell into the grid while taking account of the need for grid stability. Energy efficiency improvements in a whole range of household appliances will save consumers money, thereby attenuating fuel poverty. Fairness has been addressed both in the establishment of the 2020 renewable energy targets using the criteria of GDP *per capita*, but fairness is also addressed more widely by leaving Member States to develop the energy and climate plans at the national level that are adapted to their specific circumstances.

The multiple policy initiatives that the EU has collectively been adopting, either through joint action or through measures at national or local levels, have so far been effective in producing a coherent outcome, both in terms of energy and climate. Figure 6.4 highlights the extent to which the ongoing energy transition has led to a fundamental structural change in the EU's power generation: significantly less coal and lignite and considerably more renewable energy. Without such a coherent change in Europe's energy landscape, the EU's climate targets would never have been reached.

One conclusion emerges clearly from the analysis in this chapter: the EU succeeded in curbing the consumption of fossil fuels and is clearly now experiencing a gradual low-carbon transformation of its energy system. It is also clear that the EU has the potential to go much further. Along the way, there will be setbacks, "rebound" effects of energy efficiency improvements and geopolitical challenges that affect energy security. The governance system provides for a regular adjustment to these developments through the reviews of integrated, national climate and energy plans. That will allow the EU to realise the climate ambition that it subscribed to in the Paris Agreement.

EU-28 - Gross Electricity Generation, by Fuel - all fuels - 1990 - 2016 [TWh]

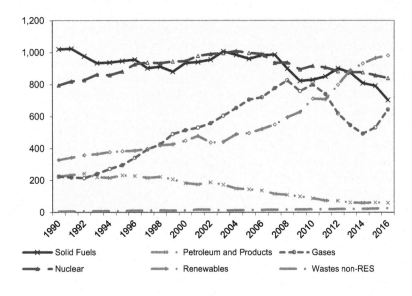

EU-28 - Gross Electricity Generation, by Fuel - Renewables - 1990 - 2016 [TWh]

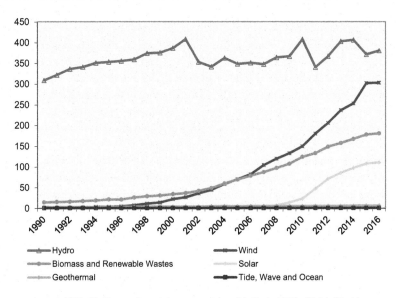

FIGURE 6.4 EU–28 Gross electricity generation by fuel, 1990–2016 (Twh)

Source: Eurostat (May 2018)

Notes

1 See p. 21 of www.consilium.europa.eu/ueDocs/cms_Data/docs/pressData/en/ec/93135.pdf.
2 Directive (EU) 2018/2001/EC of the European Parliament and of the Council of 11.4.2018 on the promotion of the use of energy from renewable sources (recast); Official Journal L328 pp. 82–209 of 21.12.2018.
3 Directive 2003/30/EC of the European Parliament and of the Council of 8.5.2003 on the promotion of the use of biofuels or other renewable fuels for transport; OJ L 123, 17.5.2003, pp. 42–46.
4 Directive 2009/28/EC of the European Parliament and of the Council of 23 April 2009 on the promotion of the use of energy from renewable sources and amending and subsequently repealing Directives 2001/77/EC and 2003/30/EC; OJ L 140, 5.6.2009, pp. 16–62.
5 Directive (EU) 2015/1513 of the European Parliament and of the Council of 9 September 2015 amending Directive 98/70/EC relating to the quality of petrol and diesel fuels and amending Directive 2009/28/EC on the promotion of the use of energy from renewable sources; OJ L 239, 15.9.2015, pp. 1–29.
6 Seven percent was the estimated installed capacity of EU biofuel production at the time.
7 Directive (EU) 2018/2001/EC of the European Parliament and of the Council of 11 December 2018 on the promotion of the use of energy from renewable sources (recast); Official Journal L328, pp. 82–209 of 21.12.2018.
8 Advanced biofuels (and biogas) made from a limited number of non-ILUC feedstocks shall constitute at least 0.2% of the blended fuel in 2022, increasing to at least 3.5% by 2030.
9 Searchinger, D. et al. (2018). "Europe's Renewable Energy Directive Poised to Harm Global Forests". www.nature.com/articles/s41467-018-06175-4.pdf.
10 Electricity Market Design: The revised legislation not yet published in Official Journal, but see: http://europa.eu/rapid/press-release_IP-18-6870_en.htm.
11 SWD (2016)410 final, Impact Assessment accompanying the Market Design Initiative proposals.
12 The ECJ however recently found that the UK capacity mechanism was not neutral and risked discrimination against demand-side response solutions.
13 Energy Efficiency Financial Institutions Group (EEFIG) Final Report, February 2015. (see https://ec.europa.eu/energy/sites/ener/files/documents/Final%20Report%20EEFIG%20v%209.1%2024022015%20clean%20FINAL%20sent.pdf) and COMMISSION/DG ECFIN, Note to the Economic Policy Committee Energy and Climate Change Working Group (19.4.2016): Investment in Energy Efficiency by Households.
14 Compared to a "business-as-usual" projection from a base year in 2005 (initially carried out in 2007), assuming continuous economic growth and no additional energy-efficiency policies above and beyond those in place in 2005.
15 Primary energy consumption measures the total energy demand of a country. It covers consumption of the energy sector itself, losses during transformation (for example, from coal or gas into electricity) and distribution of energy and the final consumption by end users. Final energy consumption is the total energy consumed by end users, such as households, industry and agriculture. It is the

energy that reaches the final consumer's door and excludes that which is used by the energy sector itself. *(Eurostat definitions)*.

16 For more details see the European Commission's Energy Efficiency Progress Report for 2018 (COM(2019)224 final of 9.4.2019).

17 Directive 2009/125/EC of the European Parliament and of the Council of 21 October 2009 establishing a framework for the setting of ecodesign requirements for energy-related products; OJ L 285, 31.10.2009, pp. 10–35.

18 Regulation (EU) 2017/1369 of the European Parliament and of the Council of 4 July 2017 setting a framework for energy labelling and repealing Directive 2010/30/EU, OJ L 198, 28.7.2017, pp. 1–23.

19 Directive (EU) 2018/844 of the European Parliament and of the Council of 30 May 2018 amending Directive 2010/31/EU on the energy performance of buildings and Directive 2012/27/EU on energy efficiency; OJ L 156, 19.6.2018, pp. 75–91.

20 Directive (EU) 2018/2002 of the European Parliament and of the Council of 11 December 2018 amending Directive 2012/27/EU on energy efficiency. Official Journal L328, pp. 210–230 published on 21.12.2018.

21 Communication "Ecodesign Work Plan 2016–2019". COM(2016)773 of 30.11.2016.

22 Listed in the "Ecodesign Working Plan 2016–2019".

23 European Commission website announcing the launch by the European Investment Bank of the Smart Finance for Smart Buildings initiative on 7 February 2018: https://ec.europa.eu/info/news/smart-finance-smart-buildings-investing-energy-efficiency-buildings-2018-feb-07_en.

24 The De-risking Energy Efficiency Platform (DEEP) is an open-source initiative to scale-up energy efficiency investments in Europe through the improved sharing and transparent analysis of existing projects in buildings and industry. More information is available on the DEEP website: https://deep.eefig.eu/.

25 Directive (EU) 2018/2002 of the European Parliament and of the Council of 11 December 2018 amending Directive 2012/27/EU on energy efficiency. Official Journal L328, pp. 210–230 published on 21.12.2018.

26 EC (2012c): Non paper of the services of the European Commission on energy efficiency Directive, Informal Energy Council, 19–20.4.2012. http://ec.europa.eu/energy/en/content/non-paper-energy-efficiency-directive (Consulted 21.4.2015).

27 Regulation (EU) 2018/1999 of the European Parliament and of the Council of 11 December 2018 on the Governance of the Energy Union and Climate Action. Official Journal L328, pp. 1–77 published on 21.12.2018.

28 Regulation (EU) No 525/2013 of the European Parliament and of the Council of 21 May 2013 on a mechanism for monitoring and reporting greenhouse gas emissions and for reporting other information at national and Union level relevant to climate change and repealing Decision No 280/2004/EC (OJ L 165, 18.6.2013, pp. 13–40) continues to apply until 2021, after which it will be replaced by the Energy Union Governance Regulation.

29 Drafts were due by December 2018; the European Commission will review the drafts by June 2019, and final plans are then submitted by December 2019.

30 Summarise the five dimensions of the Energy Union. https://ec.europa.eu/commission/priorities/energy-union-and-climate_en.

7

TRANSPORT EMISSIONS FROM ROAD, AVIATION AND SHIPPING

Damien Meadows, Alex Paquot and Peter Vis

Introduction

Transport is proving to be one of the hardest sectors to decarbonise. Today transport accounts for 21.5% of the European Union's greenhouse gas emissions,[1] increasing to more than one-quarter when aviation and maritime emissions are taken into account.[2] In Chapter 1, it was indicated that since 1990, emissions from all sectors have been reduced, with the exception of transport, where emissions have significantly increased, by 28% between 1990 and 2017 based on preliminary estimations by the European Environment Agency. The decoupling of emissions from economic growth, which is very apparent for the economy as a whole, has not happened in the road transport sector, and still less for aviation, which has more than doubled its emissions since 1990.

Transport is embedded into our way of life and the backbone of our economy, and there are no signs of mobility demand falling. On the contrary, as standards of living improve, people tend to travel further and more frequently than ever. Economists have known for a long time that the income elasticity of transport demand is higher than one. Yet we will not be able to deliver on the goals of the Paris Agreement if society is unable to develop sustainable transport with much lower emissions of greenhouse gases. That is why much stronger policy intervention seems unavoidable.

7.1 The EU's overall policy towards internalisation of external costs

In 1995, the Commission published a Green Paper, "Towards Fair and Efficient Pricing in Transport: Policy options for internalising the external costs of transport in the European Union."[3] It argued that the most obvious way forward is to make prices for transport cover the costs of all the "externalities" caused by transport. Such externalities include all adverse effects on health and the environment, congestion and the costs of accidents that are not borne directly by the transport user. The thrust was to make cost internalisation "an essential component of a multi-faceted transport strategy."

However, the emissions from transport continued to increase instead of fall, and the Commission realised that another level of ambition was necessary (see Figure 7.1). In 2011, the Commission proposed a reduction of transport emissions by 60% by 2050[4] and reconfirmed this in 2016 in its Low-emission Mobility Strategy.[5] There are multiple main elements of the strategy, including seeking higher efficiency of the transport system, low-emission alternative energy for transport and low- to zero-emission vehicles. Highlighted are the use of advanced biofuels, hydrogen and renewable synthetic fuels, removing obstacles to the electrification of transport, modal shift and smart pricing. A variety of actors have important roles to play, not least cities and local authorities.

The complex reality is that today we have an extensive mix of instruments at EU level. Several relate to transport pricing, such as the regulation of airport charges, infrastructure charging for rail, port charges for shipping, minimum excise duties on road transport fuels and road pricing for Heavy Duty Vehicles. Other instruments focus on infrastructure, such as the Alternative Fuel Infrastructure Directive and EU funding programmes, such as the Connecting Europe Facility, make considerable efforts to scale-up investments in the infrastructure needed for low-emission transport. While these instruments are far from sufficient, when taken together they do make a positive difference.

The real test of success, however, is whether emissions and environmental impacts from transport go down in absolute rather than relative terms. So far, this has not happened in a sustained way. Policies inevitably require time to show results, and changes for the better may be underway. This chapter cannot be comprehensive but reviews a number of key policies where the EU has been particularly active, such as on CO_2 standards for road vehicles and on international modes of transport, namely aviation and shipping.

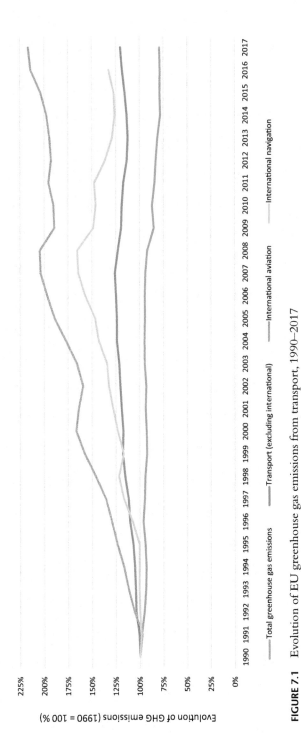

FIGURE 7.1 Evolution of EU greenhouse gas emissions from transport, 1990–2017

Source European Environment Agency (EEA): TERM Briefing, 2018[26]

7.2 Emissions from road transport

Road transport represents approximately three quarters of the emissions from transport, which is roughly one-fifth of total greenhouse gas emissions in the EU.[7] Obviously, that is a primary target for climate action. In addition, air quality is also a major concern to citizens, as the transport sector emits roughly half of the EU's NO_x emissions,[8,9] and road transport has been at the heart of the so-called "dieselgate" scandal around the illegal use of "defeat devices."

Member States have the primary responsibility for reducing road transport emissions. Despite the fact that no binding target has been set for transport, it constitutes the most important part of the mandatory target of each Member State under the Effort Sharing Regulation.[10] Interesting to note is that the power sector falls under the harmonised EU ETS; hence, the electricity consumed by trains, trams or electric vehicles is indirectly covered by this system. If the share of electric propulsion grows in the future, as is expected, regulation will shift from the level of Member States to the EU level under the EU ETS.

Fuel taxation is obviously an important incentive towards an efficient use of energy in cars, vans and lorries, in particular in comparison to other parts of the world. However, the EU only regulates the minimum levels of excise duties of mineral oil products. Product standards for transport fuels or on biofuel sustainability are also relevant for cars, vans and lorries, but their benefits have so far been very modest.

7.2.1 Regulating CO_2 emissions from cars and vans

Following agreement on the Kyoto Protocol, a Voluntary Agreement with car manufacturers was concluded in 1998. It would reduce the average emissions of new passenger cars to 140 grammes of CO_2 per kilometre (gCO_2/km) by 2008/2009.[11] This Voluntary Agreement failed, and as of 2009, emissions standards have been set in binding legislation.[12] All new passenger cars registered in the EU in 2015 and 2021 shall emit on average not more than 130 and 95 gCO_2/km respectively. In 2019, new targets were adopted for 2025 and 2030, which are respectively 15% and 37.5% lower than the 2021 target.

Car manufacturers have been meeting their 2015 obligations by the due date and the average emissions level of a new car sold in 2016 was 118.1 gCO_2/km. However, in 2017, average emissions increased year-to-year, to 118.5 gCO_2/km for the first time.

As shown in Table 7.1, the standards set will require emissions to be reduced further towards 2030, which will require intensive technological changes to the car of tomorrow. In addition to the CO_2 standards for manufacturers, an informed choice by the consumer needs to be facilitated. A mandatory label already indicates a car's fuel consumption and CO_2 emissions[13] in the showroom and in advertising material. Consumers stand to benefit from continued reduction in fuel consumption as a result of the CO_2 Regulation. For example, on average the 2021 CO_2 standard equates to approximately 4.1 litres/100 km for petrol and 3.6 litres/100 km for diesel. These savings help the consumer finance the higher upfront cost of a low-carbon car.

EU legislation with binding CO_2 targets is also in place for light commercial vehicles (vans).[14] CO_2 emissions from new vans are limited to a fleet average of 175 gCO_2/km by 2017 and 147 gCO_2/km by 2020. These targets represent reductions of 3% and 19% respectively, compared with the 2012 average of 180g CO_2/km. This corresponds to an average fuel consumption of 5.5 litre/100 km for diesel-fuelled vans by 2020. The new targets adopted in 2019 will require average van emissions in 2025 and 2030 to be 15%, respectively, 31% lower than the 2020 target.

A number of elements facilitate compliance with the legislation. The 95 gCO_2/km target in 2021 for cars allows for the use of so-called "super credits," which incentivise cars with emissions below 50 gCO_2/km, such as electric or plug-in hybrid cars. Such low-emitting cars will be counted as two vehicles in 2020, 1.67 in 2021, 1.33 in 2022 and as one vehicle from 2023 onwards. This should encourage the deployment of new technologies that

TABLE 7.1 CO_2 targets and average emissions for cars and vans (gCO_2/km, NEDC)

	Cars	Vans
1995	186	
2000	172.2	
2007	158.7	
2012	132.2	180.2
2015	119.5 – Target: 130	168.3
2016	118.1	163.7
2017	118.5	156.1 – Target: 175
2020	–	Target: 147
2021	Target: 95	–
2025	Target: 80.8 (approx.)	Target: 125 (approx.)
2030	Target: 59.4 (approx.)	Target: 101.4 (approx.)

could help realise future reductions. The contribution of these super credits however cannot contribute more than 7.5g gCO_2/km for each car manufacturer between 2020 and 2022. A similar facilitative element include eco-innovations, providing manufacturers a bonus of maximum seven gCO_2/km for CO_2 reductions through the application of innovative technologies not directly related to the engine performance.

Technological neutrality is a key principle, which leaves the choice of technologies to car manufacturers. The legislation defines for each car manufacturer a precise CO_2 target, differentiated according to the average mass of the fleet of passenger cars produced by the manufacturer. Therefore, heavier cars can emit slightly more than lighter cars, but the average of all new cars sold for each manufacturer must meet the set overall target. Historically, as is the case today, car manufacturers have chosen different technological routes, innovating with a variety of technologies. For example, as diesel engines tend to be more fuel-efficient, many of them switched to this technology in the 1990s. However, this increase in the share of diesel vehicles led to increasing problems with air quality, in particular in urban areas.

The CO_2 standards are based on the type-approval process; hence, they can only be as good as the values coming from the underlying test procedure. In recent years, there has been evidence of growing discrepancies between test cycle results and emissions in real driving conditions. There is also increasing media and regulatory interest around the use of "defeat devices" for air pollutants, or engine management systems, that car manufacturer were hiding in their cars. In view of restoring consumer confidence, test procedures have been strengthened both in terms of governance and measurement controls.

The performance of the vehicles is now measured according to a new regulatory test procedure carried out in laboratories. The type approval legislation of 2017 introduces the World Harmonised Light Vehicles Test Procedure (WLTP), developed in the context of the United Nations Economic Committee for Europe (UNECE). This replaces the old procedure known as NEDC (standing for New European Driving Cycle) which had been designed in the 1980s. The new procedure is more representative of real world driving in different conditions and reduces the risk of a creative use of the flexibilities that earlier legislation permitted.

In addition, a new market surveillance mechanism will improve the reliability and trustworthiness of the system.[15] Real world fuel consumption data will be collected and made public thanks to a standardised "on-board fuel consumption monitoring device" that will have to be installed in all new vehicles from 2021 on. Moreover, any significant deviations found during the verification of vehicle emissions in-service with the emissions determined in type-approval will be taken into account in the calculation of the average specific emissions of a manufacturer. The verification will also have to investigate the presence of any strategies that would artificially improve the performance of a vehicle during the type approval procedure.

Furthermore, penalties are part of the overall compliance provisions and remain strict. If a manufacturer's average emissions exceed its specific emissions target, the manufacturer will have to pay an excess-emissions premium equal to €95 for each gCO_2/km above its target and for each new vehicle registered in that year.

A final element of the new legislation for the period after 2020 will be a new type of incentives for zero- and low emissions vehicles. Questions were raised about the risk of losing the EU's competitive advantage in the manufacturing of cars and vans due to insufficient innovation in low-emission automotive technologies. Past CO_2 standards helped EU manufacturers to have a first mover competitive advantage at global level. However, other countries have progressively implemented their own fuel standards. Major non-EU car markets (e.g., China, California and other US States) have considered or are introducing ambitious policies, in particular to address air pollution.

It is notable that despite these "unilateral" policies developing independently from each other, today 75% of global car sales are subject to CO_2 or energy efficiency legislation. The stringency of these policies is increasing over time and converging (Table 7.1). In the US, the Californian "Zero Emission Vehicle" (ZEV) standards to support the market deployment of battery electric, plug-in hybrid, and fuel cell vehicles have been adopted by nine other States, and almost one third of new cars sales in the US take place in these ten States.[16] In China, new mandatory "new energy vehicle" (NEV) requirements covering battery electric, plug-in hybrid and fuel cell vehicles apply to car manufacturers from 2019,[17] applicable to all manufacturers with an annual production or import volume of 30,000 or more cars.

The new EU legislation to be adopted in 2019 once again includes a crediting system to speed up the uptake of zero- and low-emission vehicles (ZEV and LEV), which are defined as having CO_2 emissions between zero and 50 g/km. A manufacturer's specific CO_2 emissions target will be adjusted in case the share of zero- and low-emission vehicles in its fleet exceeds the benchmarks of 15% in 2025 and 35% (for cars) or 30% (for vans) in 2030. If a manufacturer exceeds the set benchmark by one percentage point, he will benefit from a 1% less stringent CO_2 target. This is allowed up to 5% of the target. For calculating that share, account is taken of the emissions of the zero- and low-emission vehicles, meaning that zero-emission vehicles are counted more than those with higher emissions.

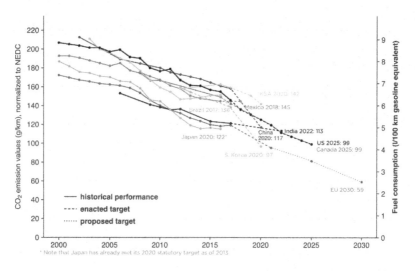

FIGURE 7.2 Average emission standards for new passenger cars (US and Canada values are for light duty vehicles)

Source: ICCT

The new crediting system should provide a strong and credible signal for the deployment of the cleanest and most efficient vehicles, while still allowing for the further improvement of the efficiency of the conventional internal combustion engines. This should yield high benefits for consumers, competitiveness and the environment. By not introducing technology specific quotas or mandates, the new legislation remains technology neutral and more cost-effective, allowing manufacturers to decide which technologies they wish to use to meet their specific emissions target.

Conclusion: Passenger cars and vans are subject to CO_2 emissions performance standards, which are being significantly tightened towards 2030. The approach is technologically neutral: no mandates, or targets, are set for specific types of propulsion technology, such as for electric cars, but strong incentives are nevertheless foreseen. The test-cycle and the enforcement mechanisms have been strengthened significantly in the light of the increasing divergence between test and real-world emissions as well as the "diesel-gate" scandal.

7.2.2 Emissions from heavy-duty vehicles (HDV) such as lorries and buses

As for passenger cars and vans, the emissions from heavy-duty vehicles, which represent a quarter of road transport emissions, continue to rise. In the period 1990–2015, they increased by around 19%[18] and another 10% increase is expected between 2010 and 2030.[19] Until recently, no specific CO_2 policy was developed for HDVs. The claim made by road haulage companies was that they already do everything to keep down the fuel consumption of their fleet, as this is a substantial part of their overall operating costs.

However, demand for freight transport follows economic growth. It is therefore not surprising that emissions keep increasing if no mitigation measures are undertaken. Moreover, a number of market barriers have limited the adoption of emission reduction measures. Few transport companies have objective data to evaluate the fuel efficiency of new HDV before purchasing them. Split incentives exist between the owners of the vehicles such as leasing companies and the operators who would benefit from lower operating fuel costs. Furthermore, HDVs are not as standardised as passenger cars and vans, which makes the monitoring of the fleet emissions more complex.

To help overcome these barriers the knowledge of the actual CO_2 emissions and fuel consumption of new HDVs needs to be improved. The Commission, in close collaboration with stakeholders, developed a simulation software, the Vehicle Energy Consumption Calculation Tool (VECTO),[20] to calculate fuel consumption and CO_2 emissions of new HDVs in a comparable manner across all manufacturers for different vehicle types. Under the type-approval framework, a so-called "certification" regulation was adopted in 2017 amongst different vehicles[21] to define the methodology each manufacturer has to use for calculating the CO_2 emissions and fuel consumption of new HDVs with the use of VECTO.

These efforts also allow the monitoring and reporting of the CO_2 emission and fuel consumption of the sector as a whole and the data will be publicly available as of 2020, based on data for 2019. Equally, the certification of HDVs will be of great importance for Member States who want to differentiate their road charging schemes according to the CO_2 performance, as would be possible through amendment of the "Eurovignette Directive" in 2017,[22] which also allows for a widening of the system to encompass all vehicles.

This groundwork made it possible for the Commission to take the next step and propose the first EU CO_2 emission standards for HDVs in May 2018, adopted in 2019 by the European Parliament and the Council.

The specific CO_2 emissions of the EU fleet of new HDVs will have to be reduced by 15% in 2025 and 30% in 2030 compared to the emissions in the reference period, which is between 1 July 2019 and 30 June 2020. As a first step, these emission standards cover the largest vehicles accounting for 70% of the total CO_2 emissions from heavy-duty vehicles.

For manufacturers failing to comply with their specific emission targets, the level of penalties are set at € 4,250 per gramme of CO_2 per tonne kilometre (gCO_2/tkm) in 2025 and € 6,800 per gCO_2/tkm in 2030.

The new legislation also includes a crediting system to incentivise the uptake of zero- and low-emission trucks. There are currently hardly any such vehicles on the road. The incentive scheme aims at stimulating investment in all segments of the fleets, including the smaller trucks used for regional delivery purposes that are most likely to be electrified first.

To reward early action, a super-credits scheme applies from 2019 until 2024. The credits gained can be used to comply with the target in 2025. A multiplier of two applies for zero-emission vehicles and a multiplier between one and two applies for low-emission vehicles, depending on their CO_2 emissions.[23] An overall cap of 3% is set on the use of super-credits in order to preserve the environmental integrity of the system. From 2025 onwards, the super-credits system is replaced by a "bonus-only" benchmark-based crediting system, with a benchmark set at 2%. The 2030 benchmark level will have to be set in the context of the 2022 review.

This means that the average specific CO_2 emissions of a manufacturer are adjusted downwards if the share of zero-emission vehicles in its entire fleet of new HDVs exceeds the 2% benchmark. The CO_2 emissions decrease is capped at a maximum of 3%.

The legislation also contains flexibilities to ensure a cost-effective implementation of the standards. In particular, a "banking and borrowing" mechanism will allow manufacturers to balance under-achievement in one year by an overachievement in another year.

In order to incentivise early emission reductions, credits can be acquired already from 2019 to 2024. The emission credits acquired in the period from 2019 to 2024 can be used to comply with the 2025 target. From 2025 to 2029, borrowing is possible, but the credits borrowed must be cleared at the latest in 2029.

Several elements reinforce the effectiveness and the robustness of the legislation, such as verification of CO_2 emissions of vehicles in service and measures to ensure that the certification procedure yields results that are representative of real-world CO_2 emissions.

Finally, an extensive review is foreseen in 2022. This review will cover, *inter alia*, the 2030 target and possible targets for 2035 and 2040, with the inclusion of other types of heavy-duty vehicles, in particular buses, coaches and trailers.

> Conclusion: As of 2019, heavy-duty vehicles (HDVs) will be subject to certification before being put on the market. A CO_2 emissions standard has been set for 2025 and 2030 for larger lorries. Legislation will enable Member States to differentiate their road charging schemes according to the CO_2 performance of HDVs, which they can also extend to passenger cars.

7.3 Emissions from international aviation and shipping

Aviation and shipping's combined climate impacts already make up around 10% of Europe's contribution to climate change, and they will grow to more than one third by 2050 unless significant measures are put in place.

The Kyoto Protocol foresees that Annex I Parties[24] should pursue limitation or reduction of international aviation and shipping emissions, working through the International Civil Aviation Organisation (ICAO) and International Maritime Organisation (IMO) respectively. The Paris Agreement did not give a new mandate to either ICAO or the IMO, but it is explicit that it covers all sources of emissions.

The EU and its Member States have constantly been at the forefront of pressing for progress in ICAO and IMO and continue to rely on them delivering action. However, taking account of scientific evidence and the disappointing contributions by ICAO and IMO to date, many argue that unilateral measures should also be taken. The EU has included aviation emissions in setting the level of ambition of its 2020 and 2030 targets.[25]

7.3.1 Aviation emissions and EU policy action

Annual global CO_2 emissions from aviation have increased from 435 million tonnes in 1990 to over 900 million tonnes by 2018. ICAO, in its 2016 Environment Report, suggested that without additional policy measures, international civil aviation's CO_2 emissions could multiply five-fold by 2050, compared to their level in 2005. Although there are many uncertainties around such illustrative modelling, we can be confident that there will be some technological and air traffic management improvements during this period. Nevertheless, ICAO itself acknowledges that considerable policy intervention will be needed to stabilise and then reduce the sector's CO_2 emissions. Let it not be forgotten that aviation also has impacts on the climate through releases of nitrogen oxides, water vapour and particles. The IPCC has estimated that the total climate impact of aviation is currently two to four times higher than the effect of its carbon dioxide emissions.[26]

To date, the majority of aviation activity has been between developed countries. In the coming decades, the aviation industry expects the majority of emissions growth to coming from flight routes to, from and between developing countries. The global differences in aviation emissions are one reason why the UNFCCC's issue of Common but Differentiated Responsibilities (CBDR) has been central to discussions. This is illustrated by Figure 7.3, showing the distribution of aviation emissions between four key regions. The US is responsible for more than five times the jet fuel use of the average global citizen, while China and India *per capita* used only 10% and 2.4% respectively as much jet fuel as the US.

Addressing the greenhouse gas emissions from aviation requires coherent policy action for international and domestic flights. International flights are covered by the mandate of the Convention on International Civil Aviation[28]

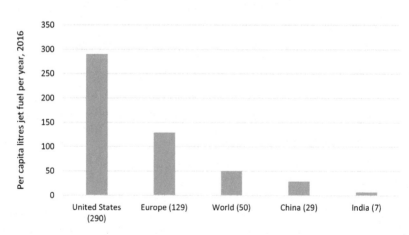

FIGURE 7.3 Average annual *per capita* jet fuel use in litres, 2016.

Source: ICCT.[27] Figure based on fuel sales per country data from US Energy Information Administration and Population based on World Development Indicators, November 2018

and action with respect to them is expected through ICAO. Domestic flights represent approximately 40% of global aviation emissions, amounting to around 400 million tonnes per year. Under the bottom-up approach of the Paris Agreement, developed country Parties should continue undertaking economy-wide absolute emission reduction targets, while developing country Parties are encouraged to move over time towards economy-wide emission reduction or limitation targets in the light of different national circumstances.[29]

As of 2012, the EU covers emissions from all flights between European airports under its EU ETS. The legislation applies to all aircraft operators active in the EU market equally, and today some 500 airlines are covered. This inclusion has resulted in cumulative surrendering of around over 100 million allowances from within the EU ETS' absolute cap on emissions and around 15 million international credits issued under the Kyoto Protocol, with respect to aviation's CO_2 emissions by the end of 2018. The EU has postponed coverage of flights to and from third countries for 11 years in order to give time for ICAO Contracting States to agree on and deliver an effective global market-based mechanism that ICAO has been seeking to put into place since 2013.

The EU decided to include aviation emissions in its emissions trading system because it was deemed to be the most realistic and cost-effective approach. Considerable investments are under way in research and development of aviation technology and fuels, and many improvements have already been put in place in terms of engine maintenance, winglets, navigation and logistics. Given the significant growth of the sector, however, much more is needed, and a carbon price is expected to encourage further improvements. The EU sees market-based measures as an essential part of a comprehensive and cost-effective approach, including for aviation.

Following the extension of the EU ETS to international aviation, airlines based in the US brought litigation to the European Court of Justice (ECJ). The ECJ however confirmed that the EU ETS law is compatible with international law, with the EU/US Open Skies Agreement and the Chicago Convention. The ECJ reaffirmed that States have the sovereign right to determine the conditions for admission to or departure from their territory and require all airlines to comply. The ECJ also confirmed that the EU ETS has no extraterritorial effect because no obligations are imposed in the territory of any other state, as inclusion within the EU ETS only arises when aircraft land or take off from an airport in a Member State. However, despite this Court ruling, many third country airlines continued to claim that the EU ETS was "extraterritorial."

Non-discriminatory application of the law is essential. Few business sectors are as international as aviation, and non-discrimination between aircraft operators on flight routes is crucial. Having different treatment for aircraft operators of different nationalities would distort competition between those operating on the same routes, so the EU ETS applies to airlines operating in the European market without distinction as to nationality. More than 100 commercial airlines based outside the EU conduct flights within the European Economic Area; these airlines all comply with the EU ETS. A number of airlines including Aeroflot, Saudi-Arabian Airlines and Air India paid significant fines for their intra-EEA emissions, at the rate of €100/tonne, before accepting to comply with the EU law.

A practical complement to emissions trading is passenger charges, which are in place in a number of countries including the US, India, Germany and the UK. Passenger charges do not apply directly to airlines, so the opportunity for airlines to oppose them is more limited. On the other hand, emissions trading is more effective in terms of giving incentives for efficiency because it rewards more efficient aircraft, flight efficiency and the use of high load factors. In addition, EU ETS allocations to airlines have been given out for free based on an efficiency-related benchmark, which rewards airlines that are more efficient.

Fuel taxes are also an option, although States have only generally applied this to domestic aviation and to business aviation. Airlines have generally been exempted from taxes such as VAT when they fly internationally, as well as from fuel taxes that, for example, apply to road transport. While it is not true that the 1944 Chicago Convention prevents the taxation of fuel supplied to aircraft, it is an enduring myth and there has thus far been insufficient political will to make significant changes to this tax exemption.

The EU always made clear that an agreement adopted at global level is by far its preferred outcome. The EU ETS has explicitly foreseen in terms of scope to take into account any future agreement. In case of failure, the EU has maintained the underlying full EU ETS coverage to all arriving and departing flights. The earliest date at which this could occur under the current legislation is 2024. By that time, ICAO should have delivered evidence that its global instrument can deliver results as promised. Removing aviation from the EU ETS would be a politically fraught exercise, also in the context of the Paris Agreement, unless other measures taken at the international level prove to be as effective.

7.3.2 Development of CORSIA within ICAO

ICAO's global market-based measure, which will be introduced voluntarily with effect from 2021, is called "CORSIA," which stands for "Carbon

Offsetting and Reduction Scheme for International Aviation." ICAO implements CORSIA through the adoption of a specific "Standards and Recommended Practices" (SARPs) that add Annexes to the Chicago Convention, along with implementing provisions and guidelines. With effect from 2027, CORSIA is meant to be mandatory for most countries.

A basic principle of the CORSIA scheme is that offsetting obligations should apply to all civil aircraft flying between two states that after 2021 have volunteered to be covered or, from 2027, to all flights between states.[30] CORSIA should last until 2035 and may be extended. A review of its operation will take place every three years. The offsetting covers only CO_2 emissions above the level of international civil aviation's CO_2 emissions in 2020,[31] unlike the EU ETS, which covers all CO_2 emissions from flights covered by the intra-EEA scope.

The following issues are of particular importance to the operation of CORSIA:

7.3.2.1 Equal treatment on flight routes

As noted previously in respect of the EU ETS, non-discriminatory application on flight routes is fundamental for a market-based measure to work. The 2016 Assembly Resolution emphasised that the ICAO scheme must "apply to all aircraft operators on the same routes between States with a view to minimising market distortion."[32] The experience with the EU ETS has shown how important enforcement is and in this area, the ICAO SARPS and draft implementing rules provide that states should not be able to enforce the scheme on any airline based in another country except if "mutual agreement" is given.[33] This is not encouraging in terms of avoiding distortions of competition. "Mutual agreement" is, by definition, bilateral and should not be a feature of a system that claims to be global and multilateral. This bilateral instead of multilateral approach towards enforcement will undoubtedly put the non-discriminatory approach at risk.

7.3.2.2 Governance arrangements and avoidance of "double counting"

At the heart of equivalence are robust and comparable Monitoring, Reporting and Verification (MRV) provisions, adopted by the ICAO Council in June 2018. The MRV obligations are meant to apply to the operators based in ICAO Member States that undertake international flights. Reporting should apply in respect of emissions from 1 January 2019 and runs to the end of 2020.

This MRV exercise alone will be valuable as a comprehensive data-gathering exercise that is needed to establish the baseline of emissions that will serve as the starting point to measure emissions growth of international aviation from 2020. This continuing monitoring obligation will serve as the basis for calculating the growth in subsequent years to determine the liability that will be distributed amongst participating operators. This unprecedented sectoral MRV exercise with global application will provide a much more detailed picture of the respective contributions to CO_2 emissions from international civil aviation, on the basis of which future policy decisions can be taken. For this reason alone, CORSIA is a worthwhile exercise, whatever its other weaknesses.

Under the Paris Agreement, practically all countries are now making commitments to reduce emissions. A new market for international credits should come mainly from Article 6 of the Paris Agreement, which should be operationalised in a way that rewards action going beyond national commitments and "business as usual." Ensuring the absence of double counting of reductions represented by eligible offsets under CORSIA is fundamental to the environmental integrity of the global market-based mechanism, for even inadvertent double counting would be as much of a blow to credibility as the circulation of counterfeit bank notes in an economy.

The EU, as the most steadfast supporter of multilateralism, has already made clear that it will implement the ICAO scheme through an amendment to the EU ETS. For this to happen, however, more clarity is needed about the eligible offsets, the oversight arrangements and finally about when and how other countries put in place their national provisions.[34]

7.3.2.3 Meaningful offsets

One area where ICAO could usefully build on experience of the EU ETS is in relation to international credits. As explained in Chapter 4, standards were needed to ensure that credits used within the EU ETS represented real emission reductions. It seems obvious that ICAO must ensure a clear legal basis to set environmental quality standards for credits.[35]

In the absence of such a legal basis, CORSIA is likely to accept several offsetting programmes, notwithstanding that there are already more than 7000 Clean Development Mechanism projects,[36] and there is no shortage of credits that could be issued. Recent research by the Stockholm Environment Institute[37] indicates that the supply potential from registered and implemented Clean Development Mechanism projects alone amounts to 4.7 billion tonnes. This volume could quickly come to market should a price signal emerge. This amount is well in excess of the expected offsetting need for the growth of international civil aviation through to 2035, estimated to be less than three billion tonnes of CO_2.[38] If supply exceeds demand there will be no scarcity and hence no meaningful price. It can be anticipated that most Clean Development Mechanism credits could be supplied at a price below €1 per tonne.

Europe has therefore argued that offsets eligible under CORSIA should be restricted to projects initiated from 2016 onwards. The year 2016 is the year of the ICAO Assembly Resolution relating to CORSIA and the entry into force of the Paris Agreement. Addressing the quality and quantity of the credits is critical to any meaningful environmental added value to CORSIA. This is still not resolved at the time of writing.

7.3.2.4 Reconciling "common but differentiated responsibilities"

There are very large differences between the levels of aviation emissions of countries. ICAO's aspirational goal of "carbon-neutral growth from 2020"[39] is a collective goal. It relies on the efforts of all states and airlines. Developing countries have consistently argued that reduction targets should take into account the fact that richer nations are responsible for the bulk of climate change to date and therefore should contribute greater reductions. This is a prominent theme in the reservations tabled by Argentina, Brazil, China, India, Saudi Arabia and Venezuela to the 2016 ICAO Assembly Resolution A39/3 relating to CORSIA.[40] This long list of reservations suggests that this matter is not closed.

7.3.2.5 Avoiding further delays to action

According to early studies commissioned by some airlines, an "open emissions trading scheme, fully linked to global carbon markets" would meet the requirements of the aviation industry.[41] The approach of airlines changed over time and their focus has more recently been on using ICAO's global scheme to replace the EU ETS.[42] There would be a chance of succeeding in this if ICAO were to deliver a meaningful, environmentally effective market-based measure. However, the timeline for this has been put back. The 2013 Assembly resolution[43] referred to "mechanisms for the implementation of the scheme from 2020," so as to "strive to achieve a collective medium term global aspirational goal of keeping the global net carbon emissions from international aviation from 2020 at the same level." By the time of the 2016 ICAO Assembly, the idea was adapted so that CORSIA should take effect in 2021 on an expressly voluntary basis, while routes between all major international aviation states would be covered between 2027 and 2035.[44]

Given the ability for states to oppose and/or file differences to the CORSIA "Standards and Recommended Practices," states will only actually participate if they agree to do so. ICAO has little ability to remedy in the case of non-compliance, and it will take time before all bilateral arrangements are aligned. Given the reservations that have been expressed to CORSIA,[45] considerable challenges are yet to be overcome.

This is in stark contrast to the reality that growing emissions from international civil aviation are far from being consistent with the temperature goals of the Paris Agreement. The liberalisation of the EU's single aviation market and the success of low-cost airlines has expanded the market and aviation emissions under the EU ETS are projected to increase continuously until 2030.[46] Similarly, in developing countries, aviation's emissions have been rising very rapidly as the sector expands to meet the enormous demand of the emerging economies. The EU agreed to extend the current intra-EEA scope for another six years until 2023.[47]

Carbon pricing remains a key element of the policy measures required first to slow the growth and then reduce the emissions from aviation. The EU continues to believe in the effectiveness of aviation's inclusion in the EU ETS until the added value of CORSIA has been proven. Despite ICAO's preference for a single uniform global measure applied by states to airlines based in those states, national and regional measures may be the outcome of a weak or ineffective CORSIA. Governments will have no other option than to apply other measures to aviation, such as passenger taxes, sustainable alternative fuel mandates and possibly taxes on tickets or fuel, if they are to have any hope of bringing aviation's emissions into line with the temperature goals of the Paris Agreement.

Conclusion: The EU ETS includes aviation between European airports on a non-discriminatory basis. ICAO has developed a global market-based instrument, "CORSIA," which should offset only CO_2 emissions from international aviation that exceed their level in 2020. The scheme will start on a voluntary basis in 2021 and is meant to become mandatory for most airlines by 2027.

7.4 Emissions from shipping

Annual global CO_2 emissions from shipping exceed 940 million tonnes. By 2050, the International Maritime Organisation (IMO) estimates that these could grow by between 50% and 250%, depending on economic and energy developments. The CO_2 is accompanied by emissions of black carbon that are considered to have significant climate impacts, in particular in the Arctic, where they dull the reflective properties of ice, thereby increasing heat absorption by ice, accelerating its melting. International journeys cover the majority of these emissions, with only 10–15% attributed to national "domestic" shipping.

The EU's CO_2 emissions from domestic shipping emissions have, by 2016, been reduced by 33.1% below 1990 levels,[48] but international shipping emissions "related" to the EU, i.e., from ships calling to EEA ports from third countries and ships sailing between two or more EU Member States, have continued to increase and are currently around 32.5% above 1990 levels.

Many technologies exist to retrofit and to build ships that are more efficient. Operational measures can reduce emissions and fuel costs, especially slow steaming. However, there are continuing market barriers that have limited the uptake of these emission reduction possibilities.[49]

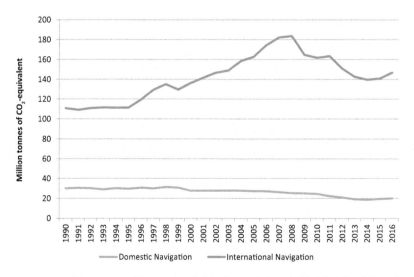

FIGURE 7.4 Domestic and international shipping emissions 1990–2016 (MtCO$_2$-eq)
Source: European Environment Agency (EEA)

In recognition of the need for economy-wide action, in 2013 the European Commission adopted a strategy for progressively integrating maritime emissions into the EU's climate policy.[50] Today the emissions of large ships using European ports are being monitored, reported and verified (MRV).[51] It has become a robust system that is a prerequisite for further action, and it already helps reduce costs by increasing information and transparency on fuel use. It also highlights the potential for more cost-effective emission reductions, including market-based measures in the future.

The EU measure also facilitated the adoption of high quality MRV standards within IMO. In 2016, the IMO's Marine Environment Protection Committee (MEPC) adopted amendments to the MARPOL Convention for a global data collection system for fuel consumption of ships.[52] The collection of fuel consumption data should start on 1 January 2019. It has a similar technical scope as the EU's MRV for shipping (with the same 5000 gross tonnage threshold), requires the same actors to report annually for their ships and introduces a document to demonstrate compliance. In early 2019, the Commission started the formal process to amend the EU MRV in order to take account of the new IMO data collection system.

In 2018, the IMO agreed for the first time a strategy for the reduction of greenhouse gases from ships.[53] It comprises both a long-term target to peak greenhouse gas emissions from international shipping as soon as possible and to reduce the total annual greenhouse emissions "by at least 50% by 2050 compared to 2008 whilst pursuing efforts towards phasing them out." There is also a commitment to improve the energy intensity of international shipping and a long list of "candidate measures" that will be further considered within the IMO. While the overall ambition of this resolution is encouraging, the reality is that agreement on any of these potential measures will be difficult. Discussions in the IMO have to reconcile the diverging principles of non-discrimination and common but differentiated responsibilities.

In the 2018 review of the EU ETS Directive, the Commission was asked to keep progress in the IMO towards an ambitious emission reduction objective "under regular review," as well as accompanying measures to ensure that the sector duly contributes to efforts needed to achieve the objectives agreed to under the Paris Agreement. Under existing rules, EU Member States can opt to include the transport sector (or parts thereof) within the EU ETS. However, no Member State has yet chosen to implement this option unilaterally, even if the EU MRV rules would considerably facilitate this, for example for short-sea shipping.

Similar to international aviation, international shipping generally benefits from tax exemption on its fuel, and no VAT applies to passenger traffic. The fact that ships have a capacity to stock a great quantity of fuel significantly limits any potential action to limit greenhouse gas emissions based on fuel sales.

The IMO has nevertheless been helpful in several areas. The introduction of the Energy Efficiency Design Index (EEDI) in 2013 is having a positive effect. More than 2,700 new ocean-going ships have been certified and the standards are becoming more stringent every five years. This does not trigger any absolute reduction of the total emissions from international shipping but improves the technology ships use.

Another area of work of the IMO relates to fuel and its sulphur content in particular, under Annex VI of the MARPOL Convention.[54] While these measures are taken in order to improve air quality, the consequential increase in fuel prices does create an increased incentive to use less fuel and consequently emit less CO_2.

The EU supports the IMO with regard to technology transfer and capacity building and makes a substantial contribution through its funding of the Global MTTC Network (GMN).[55] The aim of this Network of Maritime Technologies Cooperation Centres (MTCCs) is to help beneficiary countries limit and reduce greenhouse gas emissions from their shipping sectors through technical assistance and capacity building.

> Conclusion: The EU's system of Monitoring, Reporting and Verification of CO_2 emissions for large ships has significantly contributed to the development of a global system by the IMO. In 2018, the IMO adopted a target to reduce the emissions from international shipping by at least 50% in 2050 compared to 2008.

Conclusion

The transport sector faces a huge challenge to stabilise and then reduce its emissions of greenhouse gases. There are many useful developments across the different modes, but overall the results have been disappointing, certainly compared to the progress made in "stationary sources" in sectors such as power generation or manufacturing.

Member States are the primary level of intervention to reduce the greenhouse gas emissions of road transport. Indeed, these represent the most important part of their mandatory target for 2030 under the Effort Sharing Regulation. They have important tools at their disposal such as taxation or urban planning. This chapter only reviews a few EU measures that may be critical to help them deliver on their commitment.

In the road sector, the EU has been going through an important learning phase and is now equipped with strong tools that deliver reductions. For cars, it turned out that the voluntary approach to CO_2 performance standards was disappointing and that mandatory legislation created the right context for companies to innovate and to reduce the emissions per car. However, the "diesel-gate" scandal emphasised that solid surveillance mechanisms are also required, and these are now in place. This does not diminish the task of the Member States to bring down emissions through adapting their infrastructure, not least in readiness for the imminent arrival of many more electric vehicles. Equally, it does not underplay the critical importance of the creation of additional incentives directed towards the consumers, even if only temporary, so as to accelerate the necessary changes.

On lorries, buses and ships, EU action to define standards had to wait for the elaboration of a much more complete and solid database. The first step, therefore, consisted of generating the detailed information that is undoubtedly going to be the breeding ground for more action. For HDVs, EU standards have now been agreed for the first time for both 2025 and 2030. For shipping, the EU was successful in having a detailed and transparent MRV system put in place, as well as a data collection system adopted by the IMO. Equally, the EU continues to maintain pressure for the reduction of emissions in line with the IMO's long-term goal of at least halving shipping emissions by 2050 compared to 2008.

Contrary to the gradual steps forward in other sectors, the aviation sector has been slowing down the EU and has prevented the uptake of meaningful measures. The global offsetting instrument CORSIA had promise at the start but has been gradually undermined in its design by a lack of real commitment by states and by weak leadership from ICAO. Emissions from aviation are rising rapidly and are undermining the delivery of the goals of the Paris Agreement. The time for more tangible regional and national action is returning, as public opinion is increasingly upset about the apparent negligence of the aviation sector. This perception may help put pressure on ICAO to take climate change as seriously as it should.

Technology will undoubtedly play an important role in bringing down emissions. One of the recurring themes is that electricity has the potential to be a significant game-changer in all modes, for cars and even aviation and shipping. That is good news, as we know how to decarbonise electricity. Equally, hydrogen may find its way to ships, lorries and even planes, and fortunately we know how to produce this energy carrier in a green and sustainable manner. The EU ETS Innovation Fund could play an important role for deploying these technologies. Together with more pronounced modal shifts, these uses can deliver a critical contribution to the very urgently needed decarbonisation of the transport sector.

Notes

1 Based on data for 2017 extracted from Table A1.1 of EEA Report "Trends and Projections in Europe 2018" (First release, 26.10.2018).
2 Final Report of the High-Level Panel of the European Decarbonisation Pathways Initiative, published November 2018. Chapter 3 (p. 56): "Adding Emissions from International Aviation and Marine Navigation (in sum, 274 $MtCO_2$-eq) Increases the Full Share of Transport to 31% of Total EU-28 CO_2 Emissions". https://ec.europa.eu/info/publications/final-report-high-level-panel-european-decarbonisation-pathways-initiative_en.
3 COM(95) 691 final, 20.12.1995.
4 COM(2011)144 final, of 28.3.2011 White Paper: "Roadmap to a Single European Transport Area – Towards a Competitive and Resource Efficient Transport System".
5 COM(2016)501 final and its Annex, of 20.7.2016.
6 European Environment Agency: "Progress of EU transport sector towards its environment and climate objectives", TERM Briefing 2018 published on 22.11.2018: https://www.eea.europa.eu/themes/transport/term/term-briefing-2018.
7 EEA GHG data viewer (www.eea.europa.eu/data-and-maps/data/data-viewers/greenhouse-gases-viewer), extracted on 1.9.2017.
8 The EEA has reported that "Emissions of nitrogen oxides from international maritime transport in European waters … could be equal to land-based sources by 2020" (www.eea.europa.eu/publications/the-impact-of-international-shipping/file.
9 EEA: Air quality in Europe – 2016 report. EEA Report No 28/2016.
10 Regulation (EU) 2018/842 of the European Parliament and of the Council of 30 May 2018 on binding annual greenhouse gas emission reductions by Member States from 2021 to 2030 contributing to climate action to meet commitments under the Paris Agreement and amending Regulation (EU) No 525/2013 (OJ L 156, 19.6.2018, pp. 26–42).
11 In 1998 and 1999, the Commission signed the voluntary agreements with car manufacturing associations in Europe, Japan and Korea.
12 Regulation (EC) No 443/2009 of the European Parliament and of the Council setting emission performance standards for new passenger cars as part of the Community's integrated approach to reduce CO_2 emissions from light-duty vehicles; OJ L 140, 5.6.2009, pp. 1–25. http://eur-lex.europa.eu/legal-content/EN/TXT/PDF/?uri=CELEX:02009R0443-20130508&from=EN and Regulation (EU) No 333/2014 of the European Parliament and of the Council of 11 March 2014 amending Regulation (EC) No 443/2009 to define the modalities for reaching the 2020 target to reduce CO_2 emissions from new passenger cars; OJ L 103, 5.4.2014, pp. 15–21. http://eur-lex.europa.eu/legal-content/EN/TXT/PDF/?uri=CELEX:32014R0333&from=EN.
13 Directive 1999/94/EC of the European Parliament and of the Council relating to the availability of consumer information on fuel economy and CO_2 emissions

in respect of the marketing of new passenger cars; OJ L 12, 18.1.2000, pp. 16–23. http://eur-lex.europa.eu/legal-content/EN/TXT/PDF/?uri=CELEX:31999L0 094&from=EN.

14 Regulation (EU) No 510/2011 of the European Parliament and of the Council of 11 May 2011 setting emission performance standards for new light commercial vehicles as part of the Union's integrated approach to reduce CO_2 emissions from light-duty vehicles; OJ L 145, 31.5.2011, pp. 1–18. http://eur-lex. europa.eu/legal-content/EN/TXT/PDF/?uri=CELEX:32011R0510&from =EN; and Regulation (EU) No 253/2014 of the European Parliament and of the Council of 26 February 2014 amending Regulation (EU) No 510/2011 to define the modalities for reaching the 2020 target to reduce CO_2 emissions from new light commercial vehicles; OJ L 84, 20.3.2014, pp. 38–41. http:// eur-lex.europa.eu/legal-content/EN/TXT/PDF/?uri=CELEX:32014R0253& from=EN.

15 Scientific Advice Mechanism (SAM) (2016). Closing the Gap Between Light-duty Vehicle Real-world CO_2 Emissions and Laboratory Testing. High Level Group of Scientific Advisors, Scientific Opinion No. 1/2016. https://ec.europa.eu/ research/sam/pdf/sam_co_2_emissions_report.pdf#view=fit&pagemode=none.

16 CARB (2017) *California's Advanced Clean Cars Midterm Review – Summary Report for the Technical Analysis of the Light Duty Vehicle Standards*. www.arb.ca.gov/msprog/ acc/mtr/acc_mtr_summaryreport.pdf.

17 International Council on Clean Transportation (ICCT): Policy Update "*China's New Energy Vehicle mandate policy (final rule)*", January 2018. https:// theicct.org/sites/default/files/publications/China-NEV-mandate_ICCT-policy-update_20032018_vF-updated.pdf.

18 GHG Inventory data 2016. www.eea.europa.eu/data-and-maps/data/data-viewers/ greenhouse-gases-viewer.

19 EU Reference Scenario 2016: Energy, transport and GHG emissions – Trends to 2050.

20 For more information on VECTO, please see Annex 4 of the Impact Assessment accompanying the document Proposal for a Regulation of the European Parliament and of the Council on the monitoring and reporting of CO_2 emissions from and fuel consumption of new heavy-duty vehicles, Commission Staff Working Document Impact Assessment SWD/2017/0188 final. http://eur-lex.europa.eu/ legal-content/EN/TXT/?uri=SWD:2017:0188:FIN.

21 Proposal for a Regulation of the European Parliament and of the Council on the monitoring and reporting of CO_2 emissions from and fuel consumption of new heavy-duty vehicles, COM/2017/0279 final. http://eur-lex.europa.eu/ legal-content/EN/TXT/?uri=CELEX:52017PC0279.

22 Proposal for a Directive of the European Parliament and of the Council amending Directive 1999/62/EC on the charging of heavy goods vehicles for the use of certain infrastructures COM(2017)275 final of 31.5.2017 (1st reading in European Parliament completed on 25 October 2018).

23 A low-emission vehicle is defined as heavy-duty vehicle with emissions below 50% of the reference CO_2 emission of the sub-group to which the vehicle belongs.

24 Annex 1 to the UNFCCC.

25 See Council document WK 2310/2017 on the scope of the EU 2030 target, March 2017.

26 IPPC (1999). "Aviation and the Global Atmosphere: Summary for Policy-Makers". Section 4.8 "Over the period from 1992 to 2050, the overall radiative forcing by aircraft (excluding that from changes in cirrus clouds) for all scenarios in this report is a factor of two to four larger than the forcing by aircraft carbon dioxide alone."

27 International Council on Clean Transportation (ICCT): Blog post published 1 April 2019, author Dan Rutherford. https://theicct.org/blog/staff/whats-the-plan-sam-aviation-emissions.

28 International Civil Aviation Organization (ICAO): "Statement from the International Civil Aviation Organization (ICAO) to the Fourth Session of the Conference of the Parties to the United Nations Framework Convention on Climate Change", November 1998. https://www.icao.int/environmental-protection/Documents/STATEMENTS/cop4.pdf.

29 Article 4(4) of the Paris Agreement. https://unfccc.int/sites/default/files/english_paris_agreement.pdf.

30 Permanent exemptions exist for flights to and from Least Developed Countries, Land-Locked Developing Countries, Small Island Developing States and States with a small share of international aviation activity, but they may choose to participate in CORSIA, even if they have no airlines of their own.

31 Calculated as the average of 2019 and 2020 emissions.

32 See paragraph 10 of ICAO Assembly Resolution 39–3. www.icao.int/environmental-protection/Documents/Resolution_A39_3.pdf.

33 See sections 1.3.2 and 1.3.3.

34 Regulation (EU) 2017/2392, recital 11 and Article 28b.

35 The need for a clear legal basis upon which to operate a Global Market-Based Mechanism is addressed in a study carried out by Pablo Mendes de Leon, Vincent Correia, Uwe Erling and Thomas Leclerc for the Directorate-General Climate Action of the European Commission on "*Possible legal arrangements to implement a global market based measure for international aviation emissions*" (December 2015).https://ec.europa.eu/clima/sites/clima/files/transport/aviation/docs/gmbm_legal_study_en.pdf.

36 UNFCCC News Release: Kyoto Protocol's Clean Development Mechanism Reaches Milestone at 7,000 Registered Projects. https://cdm.unfccc.int/CDMNews/issues/issues/I_8XM9FF99N0WN7MMK9XFBJLSX23LX8Q/viewnewsitem.html.

37 Stockholm Environment Institute Project Report 2017–02: "Using the Clean Development Mechanism for Nationally Determined Contributions and International Aviation". Lambert Schneider and Stephanie La Hoz Theurer (2017).

38 Several studies are referred to in Section 4.3 of the paper, referred to in the prior footnote.

39 In the context of the Airport Carbon Accreditation initiative and of the UNFCCC, carbon neutrality means zero emissions. In the ICAO context, it means the offsetting of emissions above a baseline of the level that international aviation emissions will be in the year 2020.

40 International Civil Aviation Organization (ICAO): "*Summary Listing of Reservations to Resolutions A39-2 and A39-3*". Resolutions A39-2 and A39-3 were adopted at the 39th ICAO Assembly in October 2016. https://www.icao.int/Meetings/a39/Documents/Resolutions/summary_en.pdf.

41 See Aviation Global Deal Group discussion paper of June 2009 on a coordinated pathway towards a global sectoral agreement for international aviation emissions. www.
theclimategroup.org/sites/default/files/archive/files/AGD_Discussion_Note_
2.0_09Jun09_FINAL_.pdf. (consulted 14.12.2018). The Aviation Global Deal
Group comprised nine major airlines based outside the US, and The Climate
Group.

42 "The implementation of CORSIA will avoid the need for existing and new
carbon pricing measures to be applied to international aviation emissions on a
regional or national basis," stated in IATA Brief on CORSIA, 6.12.2017.

43 A38–18. www.icao.int/Meetings/GLADs-2015/Documents/A38-18.pdf.

44 A38–19. www.icao.int/environmental-protection/Documents/Resolution_A39_
3.pdf.

45 Reservations from Argentina, Brazil, China, India, Russia, Saudi Arabia and Venezuela are published at www.icao.int/Meetings/a39/Pages/resolutions.aspx.

46 Section 3.2 of EEA Report 14/2018. "Trends and Projections in the EU ETS in
2018".

47 Regulation (EU) 2017/2392.

48 See www.eea.europa.eu/data-and-maps/data/data-viewers/greenhouse-gases-
viewer. It should be noted that "domestic" emissions are considered for the purposes of UNFCCC reporting to be emissions internal to each Member State, as
no decision on the allocation of other emissions has yet taken place.

49 See, for example, the analysis of market barriers to cost effective maritime emission
reductions. https://ec.europa.eu/clima/sites/clima/files/transport/shipping/docs/
market_barriers_2012_en.pdf.

50 COM 2013/479.

51 Applicable since 1 January 2018 for vessels larger than 5,000 gross tonnage calling
at any EU and EFTA (Norway and Iceland) port.

52 Resolution MEPC 278(70).

53 International Maritime Organisation (IMO): Resolution MEPC.304(72) adopted
on 13 April 2018: "*Initial IMO Strategy on reduction of GHG emissions from ships*".
http://www.imo.org/en/KnowledgeCentre/IndexofIMOResolutions/Marine-
Environment-Protection-Committee-(MEPC)/Documents/MEPC.304(72).pdf.

54 International Convention for the Prevention of Pollution from Ships (MAR-
POL) of 1973.

55 Project funded by the European Union (2016–2019) entitled "Capacity Building for Climate Mitigation in the Maritime Shipping Industry," amounting to
€10 million of funding over the four-year period.

8

AGRICULTURE AND FORESTRY IN THE EU'S 2030 CLIMATE TARGET

Artur Runge-Metzger and Peter Wehrheim

Introduction

The Paris Agreement, adopted in December 2015, defined the commitment of limiting global average temperature to well below 2 °C compared to pre-industrial levels and of pursuing efforts towards 1.5 °C. It also stated that reaching this target requires "a balance between anthropogenic emissions by sources *and removals by sinks* of greenhouse gases in the second half of the century." Moreover, it requests Parties to take action to conserve and enhance, as appropriate, sinks and reservoirs of greenhouse gases, including those in forests.[1]

The European Union committed itself to the target of reducing greenhouse gas emissions by at least 40% by 2030 compared to 1990. This target is economy-wide, and agriculture and forestry will have to play their part. Both sectors not only generate emissions but also have, unlike most other productive sectors, the potential to capture and remove carbon dioxide from the atmosphere. This is not only important in view of reaching the EU target but also pivotal on a global scale as together agriculture and forestry account for about 20–25% of global emissions, mainly through tropical deforestation.[2]

In 2017, the EU adopted legislation to enhance net carbon sequestration from agricultural land and forestry inside the EU. This chapter sets out how these sectors are included within the EU's commitment for 2030, taking into account the complexity of the associated greenhouse gas cycles. Similarly, it indicates how the future Common Agricultural Policy can support the EU's climate and energy objectives.[3]

8.1 Agriculture, soils and forestry in different pillars of the EU policy

Greenhouse gas cycles related to agriculture and forestry covering both non-CO_2 and CO_2 emissions are more complex than those in many other sectors (see Figure 8.1). The two main non-CO_2 greenhouse gases from agriculture are nitrous oxide (N_2O) from the use of nitrogen fertilisers and methane (CH_4) produced by cows and sheep when digesting their feed. In 2015, these accounted for more than half of total non-CO_2 emissions in the EU, representing approximately 10% of the EU's total emissions.

These emissions are not subject to a specific sub-target for agriculture, neither at EU nor at Member State levels. Together with the greenhouse gas emissions from households, transport and waste, they constitute the sectors covered by a binding target each Member State has to respect under the Effort Sharing Regulation in view of delivering a collective reduction of 30% by 2030 compared to 2005. This Regulation takes into account the specific situation of Member States, such as Ireland, Latvia and Denmark that have large shares of non-CO_2 emissions from agriculture.

In addition to reducing emissions of non-CO_2 greenhouse gases, agriculture and forestry have the potential to help the climate problem by absorbing CO_2 into agricultural soils and forests (left side in Figure 8.1). The uptake − or removal − of CO_2 is reversible. When trees are cut down or grassland is ploughed up to create arable land, the carbon stored in the trees or soil is released into the atmosphere where it contributes to the greenhouse effect. To assess the contribution of land use and forestry to climate protection, it is critical to know how the balance between emissions and removals from agriculture and forestry evolves over time. If more CO_2 is stored in trees and less CO_2 escapes from agricultural land due to better soil protection, the contribution to climate protection increases. Moreover, agriculture and forestry, if properly managed, can produce sustainable raw materials for industry, the energy and transport sectors allowing them to replace the use of fossil fuels.

The measurement and reporting of the emissions and removals from agricultural land and forestry needed to be improved before they could be included in climate mitigation commitments. The complexity does not merely stem from the fact that there are both emissions and removals and that vast areas of land and very different land use systems have to be covered. It is also a challenge to measure the fluxes of carbon in and out of natural sinks and to identify which parts of these fluxes are anthropogenic, i.e., the result of human actions and decisions, such as the decision whether to use land for

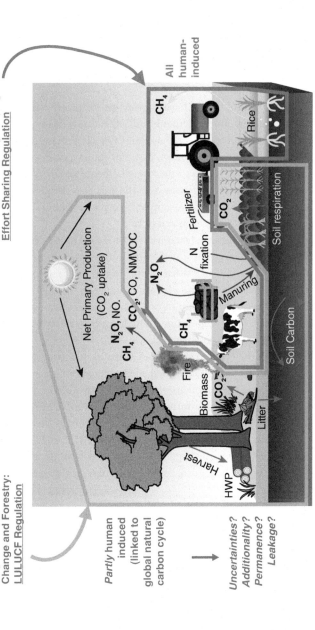

FIGURE 8.1 Land use and agriculture in EU climate policies

Source: Adapted from IPPC Guidelines 2006

settlements, agriculture or forests, or to drain peat or wetlands. Safeguards are necessary to ensure that net emissions are reduced as a result of sustainable land use management.

To address these challenges based on best available knowledge, the International Panel on Climate Change (IPCC) provided relevant recommendations[4] for the so-called "Land Use, Land Use Change and Forestry" (LULUCF) sector.

In Europe, the main LULUCF categories are cropland and grassland, wetlands, managed forestland, settlements and other lands. Figure 8.2 shows reported emissions and removals for these land use categories across the 28 Member States of the European Union. It shows the net positive contribution of Europe's forests and lands to the overall climate problem in the EU. The gross removal of carbon by forests over the past two decades was much higher than the emissions from, for example, changes in land use from forest to settlements or the emissions from arable land.

However, under the approach taken under the Kyoto Protocol, what matters in view of meeting targets is not the absolute quantities of removals or emissions from forests and land.[5] What matters are the changes in removals and emissions compared to a particular reference year. In order to identify these additional changes, the inclusion of LULUCF into national commitments are calculated against well-defined benchmarks or reference years and these are developed in the accounting rules.

Consequently, Figure 8.2 also shows the estimates of accounted removals for the LULUCF in the EU. The two shaded rectangles below the X-axis in Figure 8.2 show the accounted amounts of removals for the 15 Member States of the EU prior to 2004, during the Kyoto Protocol's first commitment period from 2008–2012 (left rectangle) and for the 28 Member States during the second commitment period from 2013–2016 (right rectangle). The accounted removals became larger in absolute terms in the second commitment period due to changes in accounting rules and due to the EU's enlargement.

Conclusion: Greenhouse emissions from agriculture and forestry are significant. They represent both emissions and removals of CO_2, and on balance they increasingly help Europe to cope with climate change.

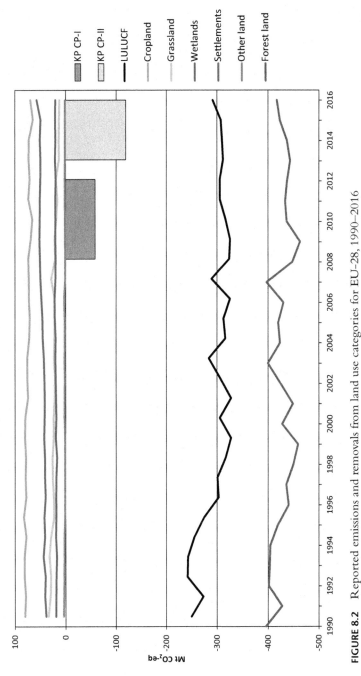

FIGURE 8.2 Reported emissions and removals from land use categories for EU-28, 1990–2016

Source: EEA Greenhouse gas – Data Viewer. UNFCCC Greenhouse Gas Inventories

Note: KP CP-I (CP-II) – the grey boxes in the figure indicate "accounted" LULUCF figures under the rules of the Kyoto Protocol for the first and second commitment periods. All other figures are "reported" figures.

8.2 The Kyoto Protocol and agriculture and land use

In 2012, the EU made a first step towards incorporating removals and emissions from LULUCF into its climate policy.[6] Accounting rules for greenhouse gas emissions and removals in the forest and agriculture sectors were established. However, this sector was still not included in the EU's domestic climate commitment from 2013 to 2020.

As a result, EU Member States improved their monitoring capacity with respect to agricultural land, not least on cropland and grazing land management. The "LULUCF Information Action" reports generated important information, for instance by identifying the most important "hot-spots" of emissions and removals and hence the most promising mitigation actions in the Member States. For the years 2013–16, LULUCF generated a surplus of removals over emissions of 100 to 120 million tonnes of CO_2-equivalent per annum with a slightly declining trend (see Figure 8.3). The main driver behind the trend is a slight decrease in the accounted sink from forests.

Figure 8.4 shows significant variation between Member States and indicates that the amount of removals realised through managed forests is sizable.[7] The accounting rules for managed forests include a provision that limits the use of removals from this category up to 3.5% of the Member States' total emissions in 1990. It also shows that results for the category of afforestation are always positive (generating accounted removals), while the results for deforestation are always negative (generating accounted emissions). Cropland sometimes creates accounted emissions (e.g., for Germany) and sometimes removals (e.g., for Portugal or United Kingdom). Some of these variations are caused by different benchmarks but are also driven by different soil types and different management practices in the Member States.

Conclusion: The implementation of the Kyoto Protocol created the first EU's accounting rules for LULUCF. Managed forests represent a considerable removal category and diminish the EU's overall net contribution to greenhouse gas emissions.

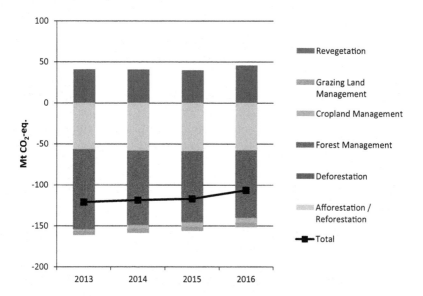

FIGURE 8.3 Estimates of accounted LULUCF emissions and removals for EU-28, 2013–2016

Source: European Commission[8]

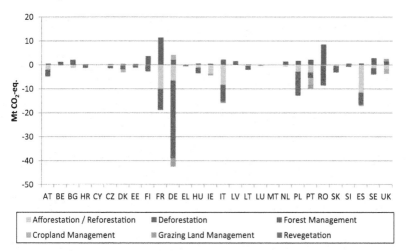

FIGURE 8.4 Estimates of accounted LULUCF emissions and removals for EU Member States, 2013–2016

Source: EEA Greenhouse gas – Data viewer. UNFCCC Greenhouse gas inventories

Note: Data on emissions and removals for activities reported under the Kyoto Protocol's second commitment period from the European Environment Agency underwent a simulated accounting process provided by the European Commission's Joint Research Centre together with DG Climate Action.

8.3 The LULUCF regulatory framework for 2021–2030

With the ratification of the Paris Agreement, the EU adopted a new 2030 target in which the LULUCF sector was included for the first time. This sector is not supposed to generate net emissions and should enhance sinks in the long term. To reach this objective, a new Regulation establishes an important new commitment for LULUCF for each Member State: the sum of emissions and removals from all land use categories must not result in net emissions, referred to as the so-called "no debit rule."[9]

In addition, removals are encouraged as a cost-effective option. If, after the use of existing flexibilities, for example trade of LULUCF "credits" between Member States (see Chapter 5), a Member State would still have emissions from LULUCF, they would have to be balanced by additional emission reductions in the other sectors covered by the Effort Sharing Regulation. If, on the contrary, a Member State generates a net removal under its LULUCF sector, it can use part of it to meet its national target under the Effort Sharing Regulation. However, for the EU as a whole the use of net removals that can be used for other sectors not covered by the EU ETS is limited to 280 million tonnes of CO_2-equivalent over the ten-year period from 2021–2030, or on average 28 million per year. This will ensure that there is still need for strong reductions in the sectors covered by the Effort Sharing Regulation. The exact amount of the limit for each Member State is determined according to its share of agricultural non-CO_2 emissions.

The LULUCF Regulation also specifies the main accounting rules for 2021–2030:

- Cropland and grassland: changes in emissions and removals in a given year are compared to the emissions/removals in an historical base-period (2005–2009), thus following the so-called "net-net" approach. The new reference period is more recent and should facilitate the identification of changes in emissions and/or removals that are the result of different management practices on agricultural land.
- Deforested and afforested land: total emissions and removals for each of the years in the period ("gross-net" approach). Afforested/deforested land normally stays in this land use category for 20 years.
- Managed forestland: accounting happens in comparison to projections of available biomass and harvest, the so-called "Forest Reference Levels." These Forest Reference Levels will take, for example, age class characteristics of the forests fully into account.

In addition, other provisions in the LULUCF Regulation ensure that the accounting of emissions and removals is fair and balanced and that adequate incentives for additional mitigation action are given:

- Forestland flexibility: Member States have some flexibility up to a specified level to increase their harvest beyond the level included in the Forest Reference Level without generating emissions. The use of this flexibility, however, is associated with safeguards: (1) LULUCF sector at EU-level must still meet the "no debit rule"; and (2) the forest sector in the Member State utilising the flexibility must, overall, generate removals.
- Harvested wood products: the legislation acknowledges the benefits of material substitution, i.e., the benefits of substituting fossil-based materials or materials production, for example construction material like steel or cement, with timber products. Emissions associated with long-lived wood products are distributed over time. For instance, emissions for harvested biomass converted into construction wood is given a half-life value of 25 years. For biomass used for the production of bioenergy, 100% of the emissions enter the accounts in the year of harvest.
- General provisions on accounting: given the complexities of monitoring emissions and removals from LULUCF, Member States are obliged to comply with best practice principles of accounting with regard to accuracy, completeness, consistency, comparability and transparency.
- Natural disturbances: exceptions are granted in events of natural disasters that are not man-made. For instance, if there is significant deforestation because of storms or due to forest fires, the associated emissions are exempt from accounting.
- Member States have to comply by 2027 and again in 2032. This will allow the calculation of the Union's total emissions and removals for each land use category compared to the average value in the period from 2000–2009.

Overall, the new Regulation develops the LULUCF sector as a new self-standing third pillar of EU climate policy, alongside the EU ETS and the Effort Sharing Regulation. It puts the focus on identifying changes in removals and emissions from land that are human induced, i.e., the result of different types of land management. The new accounting rules are built on a science-based approach, robust and close a number of loopholes that were present in the previous legislation.[10] The new LULUCF Regulation implements a major commitment of the Paris Agreement, as it incentivises conservation, restoration and expansion of forest and soil carbon sinks in view of reaching carbon neutrality by the mid-century.

Nevertheless, a reliable accounting system and methodology has some challenges.[11] For instance, the capacity of forests to absorb carbon is cyclical and depends on the age of the trees. The carbon storage will therefore increase or decrease depending on natural growth and harvest cycles under sustainable forest management. Setting a forest reference level should reflect this basic fact. The accounting system explicitly encourages the harvesting of trees at maturity, followed by replanting with young trees that will absorb the carbon.

Another major issue is the avoidance of double counting of emissions from biomass used as bioenergy. The EU's Emissions Trading System does not account for these emissions, nor does the Effort Sharing Regulation when biomass is used for heating by households. The new LULUCF Regulation now takes account of these emissions, as long as these originate from EU forests, insofar as the harvest rate is above forest reference levels. The Paris Agreement foresees equivalent rules for accounting for emissions related to biomass imported from outside the EU.

Conclusion: The EU's 2030 target includes the LULUCF (Land Use, Land Use Change and Forestry) sector and represents an emerging new third pillar in EU climate policy, in addition to the EU ETS and the Effort Sharing Regulation. This sector will not generate net emissions but enhance carbon sequestration and sinks in view of contributing to carbon neutrality by mid-century.

8.4 The enabling environment for climate action in forestry and agriculture

As outlined previously, agriculture can make a significant contribution to climate mitigation. It can reduce non-CO_2 emissions, for example in the livestock sector or in fertilisers and fuels use, and it can make an additional contribution by increasing carbon storage and sequestration. Consequently, the list of "climate smart" measures available to agriculture and forestry is much longer than in many other sectors:

- Introduction of leguminous crops on arable land;
- Reduced use of fertiliser, fuel and electricity, e.g., through use of nitrogen balances and precision farming;
- Better manure storage and spreading closer to the ground;
- Enhanced use of feed and livestock breeding and herd management;

- Improving the carbon content in agricultural soils by reducing tillage to the minimum, maintenance of permanent soil cover and enhancement of soil organic matter through mulch (conservation agriculture);
- Avoiding ploughing-up of grassland;
- Changing farming practices on former wetlands and peatlands that are rich in organic matter, the cultivation of which causes almost 100 Million tonnes of CO_2 emissions a year, i.e., almost equivalent to the entire net emission removals from LULUCF across the EU;[12]
- Combining agricultural use and trees. For example, in France, walnut trees are sometimes cultivated between fields; research in Ireland shows that planting of 200 poplar trees per hectare in a silvo-pasture system will almost treble the annual sequestration of carbon on the farm on a per-hectare basis;[13]
- Afforestation, in particular on marginal lands;
- Improving the use intensity of forests: where more trees grow or where trees grow faster, more carbon is stored on average;
- Using previously unused biomass (e.g., plant growth on roadsides but also manure, farm residues, food waste) for bio-energy production.

Many of these options also improve agricultural productivity, while at the same time contributing to the protection of biodiversity. Measures to increase the content of organic matter in intensively farmed soils can also help safeguard their long-term productivity. At the same time, such measures normally also reduce soil erosion and increase the presence of microorganisms.

Globally, natural climate solutions could provide some 37% of the cost-effective CO_2 mitigation needed through to 2030 in order to hold global warming below 2°C. One third of these would cost less than €10 per tonne of CO_2-equivalent.[14] However, trade-offs with other environmental objectives, for example if afforestation does not respect biodiversity, and with food, security objectives may occur. Therefore, a coherent approach taking into account broader sustainability considerations is necessary.

Most rural development programmes co-financed through the EU's current Common Agricultural Policy (CAP) already demonstrate "climate-smart" land management practices. The Commission proposal for the CAP in the period from 2021 to 2027 further reinforces climate objectives and measures. The proposal includes a new "green architecture" that will increase the climate ambition.[15] Through a more strategic and results-based approach, the Member States will have the flexibility to support farmers and foresters with a more targeted agricultural policy than today. Pilots for "carbon credit schemes" that would incentivise agricultural emission reductions and increases in carbon sinks could be designed and tested by individual Member States.

In addition to the Common Agricultural Policy, other EU funds support climate action related to agriculture, land use and forestry:

• The EU's Horizon 2020 programme supports a number of climate action projects related to agriculture and forestry under its "Societal Challenge 2: Food security and climate change." A specific budget of €10 billion from the new research programme "Horizon Europe" will be set aside for research and innovation in food, agriculture, rural development and the bio-economy.

• The EU's environment and climate instrument, LIFE, supports the agriculture and forestry sectors for example for maintaining or increasing biodiversity, improving air and water quality, as well as climate change mitigation and adaptation. LIFE projects are important for developing and testing new methods and knowledge related to climate-smart agriculture and land use. The LIFE programme is a "field laboratory," testing practical examples that can be rolled out at larger scale in other EU policies, such as the CAP, or national policies if tested successfully.

> Conclusion: The forestry and agriculture sectors have considerable climate action potential. Many "best practices," proven through public support, could become part of the future Common Agricultural Policy, as they are both cost-effective and enhance productivity.

8.5 Forward looking climate policies in view of 2050

Looking beyond 2030, further reductions in greenhouse gas emissions and increasing removals will become more challenging in the EU, especially where farmers and foresters experience hotter summers and scarcer water resources. While other sectors in the EU will substantially decarbonise by 2050, agricultural emissions may not reduce to the same extent and by then they may well constitute one third of total EU emissions. In the second half of the century, when global and EU emissions will have to reduce to net zero and below, the agriculture, forestry and land use sectors will be key to balancing the remaining emissions with sufficient removals.

In order to stay below 1.5°C, some authors consider that a cumulative quantity of 500 megatons of CO_2, equivalent to 115 years of the EU's 2015 greenhouse gas emissions, will have to be captured from the atmosphere by 2100.[16] One of the technologies suggested is Bioenergy Carbon Capture and Storage (BECCS). In scenarios consistent with keeping the average global temperature increase to below 2°C, the amount of BECCS could be as high as 3.3 megatons a year. However, this could require the use of 25–46% of permanently cultivated arable land.[17] All other natural climate solutions like the restoration of wetlands, reforestation, avoided forestland conversion and improved forest management will therefore have to play an equally important role.[18]

At the same time, as the global population rises, the demand for agricultural products will increase, and so will the demand for agricultural and forestry raw materials to substitute fossil fuels as feedstock. With limited arable land available, this will only be possible with very significant improvements in agricultural and forest productivity, which must also be sustainable.

Consequently, after 2030 agriculture and forestry will have to play an increasingly prominent role in EU climate policy. A more systematic monitoring of the dynamics of land use and land use changes will allow a better understanding of its drivers. Specific attention should go to potential indirect land use changes that modifications in global supply and demand of agricultural and forestry products could induce. In this respect, the EU's Copernicus satellite Earth observation programme can play a key role. It can provide the necessary geo-referenced data, especially if combined with other data sources like representative soil surveys and LULUCF inventory data.

Better modelling integrating agriculture, forestry and land use should progress considerably in the coming years. This should highlight the potential opportunities of the bio-economy in offering new productive linkages between agriculture and forestry on the one hand and industrial and energy sectors on the other. This will likely lead to the further evolution of LULUCF reporting and accounting in the coming decade, which will gradually allow for its complete integration into EU climate policy.

Finally, new policy instruments need to be developed in the coming ten years that will reward farmers and foresters for climate-smart agriculture, carbon farming and forestry and other natural climate solutions.

> Conclusion: By the mid-century, the LULUCF sectors will play a crucial role as carbon sinks to offset other unavoidable emissions. Robust monitoring and accounting systems and new policy instruments are a pre-condition to allow for the creation of adequate incentives to farmers and forest owners.

Conclusion

The agriculture and forestry sectors represent a growing area of attention for climate policy in Europe and around the world. The reason is that they are not only emitting greenhouse gas emissions but can also become an important source of removals of carbon dioxide from the atmosphere. Recent research highlights that emissions and removals from forests also play a key role in the overall pledges the signatories to the Paris climate Agreement came forward with in 2015. Globally, a quarter of the planned emission reductions by 2030 will come from the land use sector, mainly through the reduction of deforestation in developing countries.[19]

As there are still many uncertain elements about the fluxes of CO_2 the land use sector generates, much attention has been paid to how to better monitor and account for these emissions. One can expect that new technologies such as Earth monitoring and space observation will become a useful support to the statistical efforts that have so far been undertaken.

The EU started to account for these emissions as part of its implementation of the Kyoto Protocol. As these rules are about to expire, the recently adopted LULUCF Regulation will apply from 2021. At the same time, this new sector has been brought into the overall EU target of an "at least 40%" greenhouse gas reduction by 2030. Gradually a new third pillar is emerging alongside the EU Emissions Trading System and Effort Sharing Regulation.

It is now high time to develop appropriate policies to improve the uptake of carbon into Europe's soils and forests. The challenge for realising a carbon neutral Europe by 2050 will require a substantial removal capacity to neutralise emissions that are unavoidable in other sectors of the economy. All available analysis indicates that with current and expected new technologies, it will be extremely difficult to reduce the emissions by more than 80 to 90% compared to 1990, hence the need to foster the capabilities of the agriculture and forestry sectors. Local and national authorities, encouraged by European funds such as those linked to the Common Agricultural Policy, should encourage much more action in this area.

Notes

1 UNFCCC, Adoption of the Paris Agreement, Decision 1/CP.21, Paris 2015, Articles 4 and 5.
2 Overall estimates of global contributions from agriculture and forestry are associated with significant uncertainties. According to the IPCC, agriculture accounted in 2010 about 10–12% of total anthropogenic greenhouse gas emissions. The estimates for emissions associated with forests have an even wider margin and span from 10% to 20% (Zarin, D.J. "Carbon from Tropical Deforestation". Science, Vol. 336, pp. 1518–1519 (2012)).
3 European Commission Communication "The Future of Food and Farming" of November 2017.
4 IPPC's Good Practice Guidance for Land Use, Land-Use Change and Forestry Methodology Report – published 1 January 2003 by the Institute for Global Environmental Strategies (IGES) for the IPCC. www.ipcc.ch/publication/good-practice-guidance-for-land-use-land-use-change-and-forestry/.
5 Wehrheim, P., and Olesen, A.S. (2015). "Land Use, Land Use Change and Forestry – How to Enter the Climate Impact of Managing the Biospheres and Woods into the EU's Greenhouse Gas Accounting". In *Climate Change Mitigation Handbook*, eds. Van Calster, G, W. Vandenberghe and L. Reins, Edward Elgar.

6 Decision No 529/2013/EU of the European Parliament and of the Council of 21 May 2013 on accounting rules on greenhouse gas emissions and removals resulting from activities relating to land use, land-use change and forestry and on information concerning actions relating to those activities; OJ L 165, 18.6.2013, pp. 80–97.

7 All estimates are provisional and may change by 2020, for example because for the time being not all Member States account for emissions and removals for agricultural land, although this will be obligatory by 2020 (cropland/grassland).

8 Report from the Commission to the European Parliament and the Council "EU and the Paris Climate Agreement: Taking stock of progress at Katowice COP", COM(2018)716 final of 26.10.2018, p. 11.

9 Regulation (EU) 2018/841 of the European Parliament and of the Council of 30 May 2018 on the inclusion of greenhouse gas emissions and removals from land use, land use change and forestry in the 2030 climate and energy framework and amending Regulation (EU) No 525/2013 and Decision No 529/2013/EU; OJ L 156, 19.6.2018, pp. 1–25.

10 Grassi, G., Pilli, R., House, J., Federici, S., and Kurz, W.A. (2018). "Science-Based Approach for Credible Accounting of Mitigation in Managed Forests". Carbon Balance Management,Vol. 13(8).https://ec.europa.eu/jrc/en/news/measuring-climate-impact-forests-management-groundbreaking-approach.

11 LIFE News 2017, Interview with S. Kay and P. Wehrheim on the role of agriculture and forestry in the EU's climate commitments. https://life.lifevideos.eu/environment/life/news/newsarchive2017/september/index.htm#clima.

12 European Environment Agency (EEA): "Annual European Union greenhouse gas inventory 1990–2015 and inventory report 2017", chapter 6.2.7, page 734: https://www.eea.europa.eu/publications/european-union-greenhouse-gas-inventory-2017.This number is set in relation to the accounted LULUCF sink. For further explanatory reading of the importance of soil in the climate cycle, see also: European Commission (2010), "Soil: the hidden part of the carbon cycle".Brussels, ISBN 978-92-79-19269-2; https://ec.europa.eu/clima/sites/clima/files/docs/soil_and_climate_en.pdf.

13 Allen Lorcan, Farming in a Climate of Change, Dowth – driving for the world's first carbon neutral suckler beef farm. Irish Farmers Journal, 5.1.2019.

14 Griscom, B. et al. (2017). "Natural Climate Solutions". Proceedings National Academy Sciences/PNAS, 31.10.2017,Vol. 114(44), pp. 11645–11650.

15 European Commission (2019). Environmental benefits and simplification of the post-2020 CAP. https://ec.europa.eu/info/files/brochure-environmental-benefits-and-simplification-post-2020-cap_en.

16 Takeshi Kuramochi, Niklas Höhne, Michiel Schaeffer, Jasmin Cantzler, Bill Hare, Yvonne Deng, Sebastian Sterl, Markus Hagemann, Marcia Rocha, Paola Andrea Yanguas Parra, Goher-Ur-Rehman Mir, Lindee Wong, Tarik El-Laboudy, Karlien Wouters, Delphine Deryng & Kornelis Blok (2017): "Ten key short-term sectoral benchmarks to limit warming to 1.5°C", Climate Policy, ISSN: 1469-3062 (Print) 1752-7457 (Online) Journal homepage: http://www.tandfonline.com/loi/tcpo20. To link to this article: https://doi.org/10.1080/14693062.2017.1397495.

17 Smith, P., et al: "Biophysical and economic limits to negative CO_2 emissions". Nature Climate Change,Volume 6, pp 42–50 (2016).

18 Dooley, K. & Kartha, S. "Land-based negative emissions: risks for climate mitigation and impacts on sustainable development", International Environmental Agreements (2018), Volume 18, Issue 1, pp 79–98 https://doi.org/10.1007/s10784-017-9382-9.

19 Grassi, G. et al. (2017). "The Key Role of Forests in Meeting Climate Targets Requires Science for Credible Mitigation". Nature Climate Change, doi:10.1038/nclimate3227. www.carbonbrief.org/guest-post-forests-provide-quarter-paris-agreements-pledged-mitigation.

9

MAINSTREAMING CLIMATE CHANGE IN EU POLICIES

Christian Holzleitner, Philip Owen, Yvon Slingenberg and Jake Werksman

Introduction

The idea of mainstreaming climate action into other policies gained traction in the EU in the years running up to the Paris Agreement of 2015. It is increasingly recognised that mainstreaming climate change considerations into a wide range of policies is unavoidable if we want to cope with the huge challenge ahead.

There is an important opportunity in extending and reviewing environmental legislation to incorporate climate concerns. This applies in particular to the emissions of methane and black carbon that is included in EU regulations on industrial emissions. These substances, together with fluorinated gases, have a significantly stronger global warming effect than carbon dioxide. There is also the question about how to adapt the EU to the effects of climate change following from the "well below 2°C" scenario as adopted by the Paris Agreement. Member States urgently need to prepare for a ramping up of infrastructure investment responding to climate impacts. Finally, an important part of the mainstreaming relates to the EU's multiannual budget and to private finance. This is not only necessary to fund the transition required to meet the global temperature goals of the Paris Agreement but also to fund the adaptation that will be needed.

9.1 Phasing down the use of fluorinated gases

International agreements often trigger the initiation of EU policy development. In the case of the 1985 Vienna Convention for the Protection of the

Ozone Layer and the Montreal Protocol, international agreements led to the taking the first EU actions and initiatives. Similarly, methane emissions are addressed in the international context, including in cooperation with the United Nations Economic Commission for Europe's (UNECE) Convention on Long Range and Transboundary Air Pollution, the Climate and Clean Air Coalition and the Global Methane Initiative. These examples illustrate the interrelations, once again, between the compatibility of EU actions with international processes that can act as a catalyst, or at the very least as a framework, for European action.

9.1.1 Addressing the hole in the ozone layer internationally

The Montreal Protocol on ozone-depleting substances was adopted in 1987 to counter the effects of chlorofluorocarbons (CFCs) and then hydrochlorofluorocarbons (HCFCs) on the world's ozone layer. Following the phase-out of CFCs and HCFCs, hydrofluorocarbons (HFCs) became the alternative of choice for use in the refrigeration, air-conditioning and heat-pump sectors as well as being used as the extrusion agent in foams and aerosols.

These fluorinated gases, also known as "F-gases," are very powerful greenhouse gases. Their potential warming effect on the atmosphere can be up to 23,000 times higher than carbon dioxide's. Since 2006, the EU has prohibited certain uses of hydrofluorocarbons (HFCs) and other fluorinated greenhouse gases, given the availability of alternative substances. It put in place strict rules to prevent leakage of the gases from products and to ensure appropriate treatment at the end of their life.[1] In Europe, as elsewhere in the world, HFCs helped solve the problem of protecting the ozone layer but, given their high greenhouse gas potency, made the climate problem worse.

The EU was very active in remedying this situation, domestically and internationally. In 2016, in Kigali, the Parties to the Montreal Protocol agreed to a global phase-down of the production and consumption of HFCs. This Kigali Amendment entered into force in January 2019. It represents a "top-down" approach of target setting that will be facilitated by a Multilateral Fund ensuring developing country compliance.

There are three separate schedules to "phase-down," or reduce the production of, HFCs:

- Developed countries: from a 2011–2013 baseline, a reduction of 85% by 2036;
- Developing countries (other than those specified in the 3rd schedule): from a 2020–2022 baseline, with a 2024 freeze, a reduction of 80% by 2045;

• Developing countries (Bahrain, India, Iran, Iraq, Kuwait, Oman, Pakistan, Qatar, Saudi Arabia and the United Arab Emirates): from a 2024–2026 baseline, with a 2028 freeze, a reduction of 85% by 2047.

By 2050, this phase-down will avoid over 73 billion tonnes CO_2-equivalent, and at least a further 10 billion tonnes CO_2-equivalent of HFC23 by-production will be destroyed. According to the United Nations Environment Programme, this will avoid up to 0.4°C of global warming this century. It is expected to cost, in terms of the necessary replenishments of the Multilateral Fund, between $6–10 billion over the same period.

With implementation of the Kigali amendment, the "remedy" for global warming HFCs that initially replaced ozone-depleting substances (CFCs and HCFCs) is agreed. The phasing-out over time is to enable developing countries in particular to adapt gradually and reduce the costs of compliance. It has taken time to acknowledge unintended consequences of action to address the ozone hole and to address them, but while time is often needed to adapt, the result is more robust.

Conclusion: There will be 0.4°C less global warming by the end of the century thanks to the Kigali Amendment to the Montreal Protocol. It phases-down the use of HFCs and will avoid more than 80 billion tonnes of CO_2-equivalent in greenhouse gas emissions. It corrects the unintended consequences of phasing out ozone-depleting CFCs and global warming HCFCs.

9.1.2 EU legislation implementing the Montreal Protocol and the Kigali Amendment

The EU started regulating HFCs from 2006. The main refrigerant used in mobile air conditioning, HFC134a, has a global warming impact 1,300 times higher than CO_2. To pre-empt an expected sharp increase in these emissions, the 2006 EU Mobile Air Conditioning (MAC) Directive[2] required all new types of passenger cars sold from 2011 to use cooling agents with a greenhouse warming potential of less than 150 times that of CO_2. From 2018, all new passenger cars have to use these less climate potent refrigerants in their air conditioning systems. The EU Directive on handling of end-of-life vehicles[3] has required the collection and proper disposal of scrapped mobile air conditioners.

Perfluorocarbons (PFCs) released from primary aluminium production are covered by the EU's Emissions Trading System from 2013 onwards. The small number of producers in the semiconductor industry that emit PFCs made a voluntary agreement to reduce their absolute PFC emissions by 10% in 2010 compared to 1995, and they made a 41% reduction over this period.[4]

With the help of all the previously mentioned measures, European emissions of "F-gases" have stopped increasing and have stabilised at levels of 110–120 million tonnes of CO_2-equivalent. This is still, however, a level inconsistent with a 40% greenhouse gas reduction foreseen for 2030. In 2014, therefore, a new F-gas Regulation[5] was adopted to phase-down the total amount of HFCs that can be sold in the EU from 2015 to one-fifth of today's sales by 2030. This will cut emissions at marginal costs per tonne roughly equal to the overall marginal costs needed to reduce the EU's greenhouse gas emissions by 40%. By 2030, the Regulation is expected to reduce the EU's F-gas emissions by around 70 million tonnes of CO_2-equivalent, in other words, two-thirds below today's levels (see Figure 9.1).

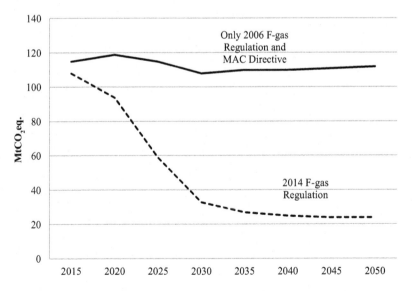

FIGURE 9.1   Expected impact of the 2014 F-gas Regulation on EU HFC emissions (in $MtCO_2eq$) to 2050

Source: European Commission

This measure is a very important piece of legislation that stimulates innovation and boosts European companies' leadership in the sector. Data from 2015, 2016 and 2017, the first three years of the phase-down,[6] show the target being achieved. In addition, the Commission has published a number of reports addressing specific issues. These include barriers caused by standards allowing low global warming potential alternatives to HFCs to come to the market,[7] availability of qualified personnel to ensure the effective installation and servicing of equipment,[8] the efficacy of the quota allocation system[9] and alternatives to HFCs for use in multipack centralised refrigeration systems.[10]

> Conclusion: The EU has put in place legislation to phase-down the use of fluorinated gases and to incentivise the deployment of alternatives with much lower global warming potential, such as CO_2 itself. This has been crucial in view of the more widespread use of air-conditioning, in light-duty passenger vehicles for example.

9.2 Short-lived climate forcers: methane and black carbon

Short-lived climate forcers are pollutants that stay in the atmosphere for a limited number of years while having a significant impact on the climate because of their high global warming potential. Apart from fluorinated gases, the most important short-lived climate forcer is methane. Another is black carbon, or soot, caused by the incomplete combustion of fossil fuels or biomass.

Methane is a greenhouse gas covered specifically by the Kyoto Protocol but black carbon is not, yet it has a global warming effect by absorbing sunlight and reducing the albedo effect of ice and snow. The EU's nationally determined contribution under the Paris Agreement applies as a metric for greenhouse gas emissions a global warming potential on a 100-year timescale, in accordance with the IPCC's Guidelines for National Greenhouse Gas Inventories. The EU's climate legislation for 2020 and for 2030 is based on the use of these values. Over a 100-year period, methane has a greenhouse gas warming potential 25 times higher than carbon dioxide.

The major sources for methane in the EU are agriculture, followed by waste management and energy. Methane emissions from agriculture arise in particular from livestock, caused by enteric fermentation in cattle and sheep, as well as during the management of animal manure. Methane emissions from the waste sector are caused by waste decomposition in landfill sites and from wastewater treatment. Energy sources of methane result from both leakage from fossil fuel extraction and production, as well as leakage from the pipeline transportation of gas.[11]

Data on methane (CH_4) emissions, for example from leakage of the oil and gas sector in the EU, are based on national reports that are reviewed by international experts and (partially) rely on data from industry. The Commission

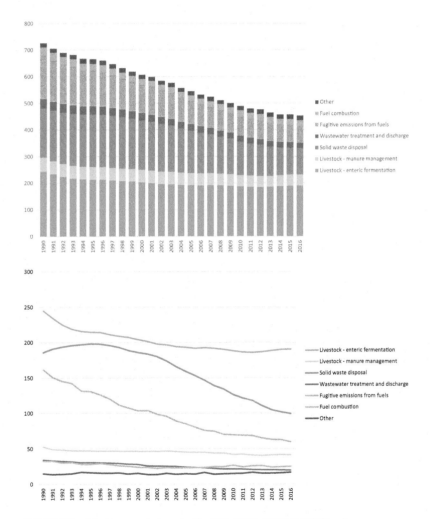

FIGURE 9.2 Methane emissions in the EU, 1990–2016 ($MtCO_2eq$)

Source: European Environment Agency

supports the development of independent CH_4 measurement tools through the EU's R&D programme "Horizon 2020." This includes the use of satellite data to understand the scale and variation of emissions from different sources better. This is expected to be operational in 2022.

As explained in Chapter 5, methane emissions from all sources in the EU are covered by the Effort Sharing Regulation. With current legislation in place, methane emissions are expected to decrease in the future (see Figure 9.3). Looking at agriculture, methane emissions from livestock are driven by demand for animal products (dairy as well as meat) and productivity changes. Despite the increase in animal numbers, methane emissions from agriculture are expected to remain stable at a level of some 250 million tonnes of CO_2-equivalent. This is mainly the result of the expected increase in the use of anaerobic digestion to recover heat and electricity. Several Member States have incentives in place to foster this technology as part of their national strategies to meet the agreed national targets for renewable energy in the EU in 2020. The Commission is also examining the possibility of funding options to reduce methane emissions in agriculture as part of the Common Agricultural Policy.

In the energy sector, methane emissions are expected to decline over time from around 90 million tonnes of CO_2-equivalent in 2005 to less than

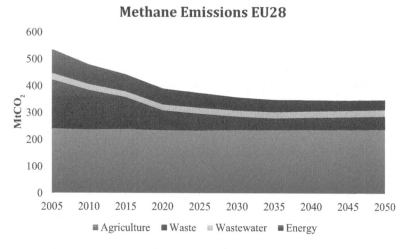

Methane Emissions EU28

FIGURE 9.3 The expected development of EU methane emission with current legislation

Source: Capros, P. et al (2016) "*EU Reference Scenario 2016: Energy, transport and GHG emission trends to 2050*", Luxemburg, Publication Office of the European Union

50 million tonnes in 2030. This decline is mainly due to the expected reduction in the production of coal and, to a smaller extent, reduced oil and gas consumption. Methane leakage from gas consumption is not expected to decline much since gas consumption is foreseen to go down only marginally and demand is expected to shift in the EU to countries with higher leakage rates, thus partially compensating for the beneficial effect of the reduction in consumption.

In the wastewater sector, methane emissions are expected to remain more or less constant through to 2050 at a level of around 40 million tonnes of CO_2-equivalent. The emissions from the waste sector as a whole are foreseen to decline significantly, following the implementation of the Landfill Directive.[12] The Landfill Directive requires the volume of biodegradable waste that is sent to landfill to be reduced by 65% in 2018. Several Member States, such as Austria, Belgium, Denmark, Germany and the Netherlands have even banned the landfilling of biodegradable waste. Methane emissions from the waste sector fell by 35% between 1990 and 2010, mainly because of the Landfill Directive.

Methane is not only a greenhouse gas but also a major ozone precursor to background ozone concentrations. Following adoption of the National Emission Ceilings Directive[13] that sets limits for air pollutants, the Commission published its First Clean Air Outlook.[14] Member States have to submit their National Air Pollution Control Programmes by 1 April 2019.

Black carbon particles enter the atmosphere around the world from sources including the burning of household firewood, other biomass and coal. Other sources are exhaust fumes from road vehicles, agricultural and industrial machinery and power plants. Sooty particles also form when oil and is burnt or flared-off in oil fields. They also form during forest fires. The World Health Organisation has reported that black carbon causes health problems including heart, circulatory and respiratory diseases, as well as premature deaths. Temperatures in the Arctic are rising more than twice as rapidly as global average temperatures. Approximately 20 to 25% of warming in the Arctic today is caused by black carbon. The sooty particles fall onto snow and ice, absorb sunlight and reduce the reflecting properties of the snow-covered surface. The black carbon particles are themselves heated by the sun, resulting in melting of snow and ice.

Regarding black carbon emissions, the Ecodesign Directive regulates the energy efficiency of a large number of products including space-heating appliances and stoves. These requirements will reduce black carbon emissions from households and smaller combustion sources in the EU by 25% in 2020 and 75% in 2050.[15] In addition, the National Emission Ceilings Directive requires the reduction of the emissions of small particles (PM2.5) in the EU by 49% by 2030 compared to 2005. It sets country specific national ceilings for reducing these emissions. The Directive also requires the prioritisation of the reduction of black carbon emissions in these reduction efforts.[16]

Conclusion: EU environmental legislation contributes to climate action. The National Emissions Ceilings Directive regulates methane, fine particulates and black carbon from large installations. The Ecodesign Directive incorporates the efficiency of space-heating appliances and stoves, thereby reducing their emissions of black carbon.

9.3 Adaptation to climate change

The Paris Agreement aims at limiting climate change to "well below 2°C." This implies that to some extent climate change is going to affect our societies. Research[17] suggests that even if all fossil fuel emissions had stopped in 2017, by 2100 we could still experience an average surface temperature warming of 1.3° C. This will require a revision of many "traditional" policies such as in the fields of infrastructure, insurance or dealing with exceptional circumstances such as forest fires or floods that are likely to happen more frequently.

9.3.1 Adaptation and the Paris Agreement

The Intergovernmental Panel on Climate Change (IPCC) defines adaptation as an "adjustment in natural or human systems in response to actual or expected climatic stimuli or their effects, which moderates harm or exploits beneficial opportunities." According to the latest IPCC Assessment Report,[18] climate change has already caused significant changes in natural systems across the world. The fact that the Arctic is warming more than twice as fast as the rest of the planet accelerates the melting of ice and exacerbates sea-level rise. Climate extremes, such as heavy precipitation, heat waves, droughts and floods, are increasing in frequency and significance. The IPCC has concluded that it is likely that droughts have been more intense and longer, in particular in southern Europe and West Africa, because of changes in climate.[19]

Anticipating the adverse effects of climate change and taking appropriate action to prevent or minimise damage caused by rising average global temperature requires a variety of measures. These include using scarce water resources more efficiently, adapting building codes to future climatic conditions and extreme weather events, building flood defences such as raising the levels of dykes, developing drought-tolerant crops, choosing tree species and forestry practices less vulnerable to temperature and precipitation change, storms and fires, and setting aside land corridors to help species migrate. The ability of society to respond depends on the level of development, the adaptive capacity and the resources available. Less developed countries and marginalised people are especially vulnerable.

The Paris Agreement recognises the importance of adaptation through the establishment of a "global goal on adaptation of enhancing adaptive capacity, strengthening resilience and reducing vulnerability."[20] The Paris Agreement introduces obligations for adaptation planning and the implementation of actions. Adaptation action is also included within the regular stock takes that will assess collective progress and increase national and regional ambition over time.

9.3.2 The EU Adaptation Strategy

While impacts vary across the EU, all Member States are vulnerable. European land temperatures were the highest on record during the last decade.[21] Precipitation patterns are changing, generally making wet regions wetter and dry regions dryer, particularly in summer; climate-related extremes such as heat waves and heavy precipitation are increasing in frequency in many regions. Mediterranean Member States in particular are likely to face severe challenges such as heat extremes, water scarcity and forest fires. The total economic losses caused by weather and other climate-related extremes in Europe for the period from 1980–2016 amounted to over €436 billion.[22]

Europe, like other parts of the world, needs to design and implement policies and measures to deal with climate impacts and their economic, environmental and social costs. Well-planned, early adaptation action saves money and lives later, by prioritising coherent and flexible approaches. Adaptation strategies are necessary at different levels of governance, such as European, national, regional authorities and municipalities, taking into account the respective roles of different actors.

The EU Strategy on adaptation to climate change[23] sets out a framework and mechanisms for taking the EU's preparedness for current and future climate impacts to a new level. Complementing the activities of Member States, the strategy promotes greater coordination, planning and information sharing and calls for incorporation of adaptation considerations in all relevant EU policies. It also ensures that the EU funds are used towards adaptation efforts.

As of April 2018, 25 Member States have adopted national strategies on adaptation. The most commonly identified vulnerable sectors across Europe are (1) agriculture and forestry, (2) environment, ecosystems and biodiversity, (3) health (plant, animal and human) and (4) water resources and management. The Joint Research Centre, the European Commission's in-house science and knowledge service, is modelling climate change impacts in Europe under different climate change scenarios, as well as exploring adaptation options in specific sectors.

The main area of European-level involvement in adaptation is funding. Since 2014, financial resources have been mobilised for climate adaptation through the LIFE programme.[24] LIFE funding can be used for adaptation activities in vulnerable areas in Europe, for activities that for example support the implementation of adaptation plans and strategies, comply with the requirements of the Covenant of Mayors for Climate and Energy or promote the use of nature-based solutions and green infrastructure as innovative adaptation solutions.

Between 2014 and 2020, EU structural and investment funds, including the European regional development funds, cohesion funds and the European Agricultural Fund for Rural Development intends to contribute €13.9 billion in direct adaptation funding.[25]

9.3.3 Adaptation at the level of towns and cities

In 2015, the "Covenant of Mayors for Climate and Energy" was launched to support the implementation of the EU's 40% greenhouse gas reduction commitment by 2030 and to follow a joint approach to tackling climate change mitigation and adaptation.[26] Over 1000 cities have committed to adaptation planning and action. The Covenant gives visibility to cities' commitments and actions, facilitates the exchange of experiences and provides technical support. Across the EU, around 40% of cities of more than 150,000 inhabitants are estimated to have adopted adaptation plans to protect their citizens from climate impacts.

Climate-ADAPT,[27] the web-based Climate Adaptation Platform, focuses on disseminating information on adaptation. It helps users access and share information on current and future vulnerability of regions and sectors, on national/transnational adaptation strategies, case studies and potential options, as well as on tools that support adaptation planning. The Climate-ADAPT database currently contains more than 1500 resources on countries adaptation policies and actions, including ecosystem/biodiversity-based adaptation strategies. The information provides a useful source of information and resources for decision-makers, and the platform could potentially inform similar initiatives in other regions of the world.

9.3.4 Incorporating adaptation into EU water policy, disaster risk reduction and CEN/CENELEC

The EU's Adaptation Strategy emphasises the importance of mainstreaming adaptation into other policy areas, such as water and disaster risk reduction (DRR).

The EU has inserted adaptation to climate change into its water policy, such as the Water Framework and Floods Directives. Several parts of Europe are already experiencing violent flash floods, with costly consequences in terms of human life and financial impacts. Between 2001 and 2010, flood damage cost nearly €5 billion each year, and by 2050, that figure could increase five-fold. In 2015, Member States finalised their Flood Risk Management Plans, which also address climate change.

The EU Adaptation Strategy calls for implementation of adaptation policies in the context of DRR. Climate change is a risk-multiplier and needs to be fully integrated into the full disaster cycle (prevention, preparedness, response and recovery).

Ensuring the resilience of infrastructure to current and projected climate impacts is critical for the sustainability of Europe's economy. In 2016, a "Guide for addressing climate change adaptation in standards" was adopted by the European Committee for Standardization (CEN) and the European Committee for Electrotechnical Standardization (CENELEC).[28] The guide is intended to help standard developers address the consequences and implications of climate change. It includes a checklist to help establish whether climate change adaptation is relevant to a particular standardisation activity and a decision tree to help identify which actions should be taken. The next steps include the revision and development of the identified standards aiming to improve the resilience of European infrastructure in three priority sectors to the adverse effects of climate change – transport infrastructure, energy infrastructure and the construction of buildings.

Working towards enhancing the adaptation capacity of European infrastructures, the European Financing Institutions Working Group on Adaptation to Climate Change published a report to help practitioners and beneficiaries ensure that climate change risks and vulnerabilities are properly assessed and integrated into project planning and design.[29] Guidelines were published on mainstreaming climate change – mitigation and adaptation – in major projects in the EU at all stages of the project cycle.[30]

9.3.5 International dimension to climate change adaptation

Helping countries adapt to climate change is not only about fairness: it is also in the EU's own interest to help build adaptation capacity elsewhere. Poor adaptation in other countries may disrupt transport infrastructures, lower their agricultural yields and force people to migrate. Climate change impacts – such as extreme weather events, changes in precipitation patterns, droughts and desertification – can have direct and indirect security impacts and interact with other features of the social, economic and political landscape, with potentially destabilising effects that need to be addressed. They can be a threat multiplier that exacerbates the risk of conflict and displacement of people.

> Conclusion: The Paris Agreement recognises that climate change is already happening, and that adaptation action is required. The EU has developed information sharing, ensures that long-term investments are more climate resilient and supports adaptation needs internationally, especially in the Small Island Developing States and Least Developed Countries.

9.4 Mainstreaming climate into the EU budget and developing sustainable finance

A critical area for climate mainstreaming is finance. The 2011 "Low-carbon roadmap to 2050" already indicated that the total investment needs of the EU would require a major increase in public and private investment.[31] Similarly, in its 2018 long-term strategy communication towards a climate-neutral Europe by 2050, the Commission confirms that additional investments of €175 to €290 billion a year will be necessary.[32] It is clear that the availability of sustainable finance is a crucial pre-condition for the success of the low-carbon transition.

9.4.1 EU public funds

With regard to public finance, the EU budget should play a major role because climate action is, alongside better protection and management of external borders, defence and security, a public good that generates a clear European added value.[33] Therefore, the EU took the responsibility of mainstreaming the climate dimension into its budget. Until 2020, a goal of 20% was decided for climate related spending in order to encourage the consideration of climate in the design and implementation of all EU spending programmes.[34]

At the same time, the mainstreaming commitments were further translated into similar goals for the European Fund for Sustainable Development (28%), the European Investment Bank (25%), the European Fund for Strategic Investment 2 (40%) and several Multilateral Development Banks, ranging from 28% to 40% climate relevance, with stricter tracking methods.

According to the EU's budget implementation review, more efforts are clearly necessary, even if to date the 20% target is close to being achieved.[35] It is set to deliver slightly more than €200 billion. The 20% mainstreaming target helped to bring the climate dimension into the discussions with all stakeholders on programme design and implementation, not least at the Member State level.

The European Court of Auditors' performance audit on climate spending in the 2014–2020 EU budget[36] treats the 20% climate objective as the core of mainstreaming and as a financial objective in itself. In May 2018, the Commission proposed that this mainstreaming of climate within the EU budget be continued and even be increased to 25% (or €320 billion) in the next Multiannual Financial Framework proposal for 2021–2027.[37] The means of achieving this target have already been included in all of the Commission's programme proposals for 2021–2027, for example the Common Agricultural Policy up to €146 billion,[38] the European Regional Development Fund up to €68 billion[39] and the research programme Horizon Europe up to €33 billion.[40]

A key priority is financial support for the development of low-carbon technologies and business models that will be necessary for reaching the long-term climate goals. Innovation offers a strategic opportunity for increasing the EU's competitiveness in relation to the rest of the world. A low-carbon economy will reduce the EU's dependence on fossil fuel imports and redirect value creation to happening from within the European economy. At the same time, European industry can take advantage of the low-carbon transition in Europe and become a leading technology provider, in particular in fast-growing Asian markets.[41] The EU has the ambition to become a leading global provider for low-carbon technologies.

The EU supports low-carbon technologies at all levels of their development from research to market uptake. As mentioned, the new research programme "Horizon Europe"[42] should spend 35 % of its budget on climate action across industry, energy, transport and the bio-economy. However, it is an additional challenge to bring innovative low-carbon technologies such as carbon capture and storage or hydrogen to the market because in many cases the regulatory environment is not yet adapted for those technologies. The current level of the carbon price does often not allow new technologies to become profitable in the short- or medium-term and the necessary infrastructure is often still lacking.

The EU has been able to gather experience of providing public support to large-scale demonstration of pre-commercial technologies through revenues from the EU Emissions Trading System. The financial support needs to be targeted towards the project's cash-flow needs and should offer a perspective to become competitive over the project's lifetime. The new Innovation Fund introduced by the latest legislative amendment to the EU ETS is likely to provide at least €10 billion of support from the revenues of the EU Emissions Trading System over the period from 2020–2030. This fund will enable flagship demonstration projects across renewable energy, carbon capture use and transport, energy storage and energy-intensive industries.

The EU's regional and infrastructure funds, which should spend at least 30% on climate action,[43] should support the rollout of low-carbon technologies that are ready for the market. In addition, funding programmes such as renewable energy auctions by Member States, as well as the Modernisation Fund, which is also financed from the revenues of the EU Emissions Trading System, should further support the low-carbon transition across all regions of the EU.

The financial support in the earlier stages of development – through Horizon Europe and the Innovation Fund – should be provided primarily by grants with a view to cover the inherently higher risks. Later, as technologies come closer to market readiness, it will be more effective to use financial instruments such as loans with a public guarantee under the InvestEU programme,[44] with a view to realise a higher leverage of public funds and draw in an increasing share of private funding.

The regional funds and the Modernisation Fund have an additional role to play because the transition to a low-carbon economy will inevitably lead to the reduction of certain economic activities. In 2015, there were, for example, 237,000 direct jobs in coal and lignite mining and power plants – the majority of them in just a few regions. By 2030, it is estimated that around 160,000 of these may be lost.[45] The transformation of those regions should not be left to Member States alone but be supported at the European level because a timely exit from fossil fuels will reduce emissions for the benefit of all EU citizens. The regional funds are a well-tested policy instrument for these purposes. The Modernisation Fund can provide additional support for a timely and just transition, for example through subsidies for reskilling of employees.

With regard to the bio-economy, at least 40% of the EU's budget for the Common Agricultural Policy should be related to climate action.[46] Farmers should comply with a basic set of standards concerning climate and environment. In addition, Member States can set up several types of incentive schemes to reward agricultural practices that are beneficial for the climate, for example payments for the maintenance of and the conversion to organic land or for other types of interventions such as agroecology, conservation agriculture and integrated production.

9.4.2 EU's international financial support

The 20% mainstreaming goal also applies to the EU's development cooperation policies spanning all sectors, with a particular focus on adaptation, building resilience, disaster risk reduction and renewable energy. In the area of adaptation, the EU's flagship initiative is the Global Climate Change Alliance (GCCA+) that currently supports 49 projects in 38 countries and is one of the world's largest climate initiatives. It has invested €285 million from 2008–2013 and has an expected commitment of around €420 million for 2014–2020. The initiative builds on two pillars: policy dialogue and technical and financial support for implementation of national climate change adaptation and mitigation policies. It will continue helping vulnerable countries, mainly Small Island Developing States and Least Developed Countries.

The EU and its Member States are collectively the largest contributors in the world when it comes to bilateral assistance for climate change. The climate finance contribution of the EU and its Member States amounted to more than €20 billion in 2017 for both mitigation and adaptation activities.

Mainstreaming climate considerations is even more important when it comes to the enormous amount of private funds that are being invested on a daily basis. That makes the concept of sustainable finance key to the delivery of the targets. Sustainable finance is the provision of finance to investments taking into account environmental, social and governance considerations. It aims to support economic growth while reducing pressures on the environment, addressing greenhouse gas emissions, tackling pollution and minimising waste, as well as improving efficiency in the use of natural resources. Projects financed need to take both climate mitigation and climate resilience into account.

9.4.3 Sustainable finance

The need for sustainable finance has gained traction following the adoption and entry into force of the Paris Agreement and specifically its Article 2(1) (c), which calls for "making financial flows consistent with a pathway towards low greenhouse gas emissions and climate resilient development."

The European Union strongly supports this ambition. In line with its September 2016 Communication on "Capital Markets Union – Accelerating Reform,"[47] the Commission established a High-Level Expert Group on Sustainable Finance[48] to advise on developing a comprehensive EU strategy on sustainable finance. This High-Level Expert Group provided recommendations and published its final report in January 2018. As a follow-up to these recommendations in March 2018, the Commission adopted an Action Plan on sustainable finance[49] that aims to:

1 Re-orient capital flows towards sustainable investment, in order to achieve sustainable and inclusive growth;
2 Manage financial risks stemming from climate change, environmental degradation and social issues;
3 Foster transparency and long-termism in financial and economic activity.

The Commission has subsequently acted on this Action Plan and in May 2018 adopted a series of related legislative proposals on a framework to facilitate sustainable investment,[50] disclosures relating to sustainable investments and sustainability risks[51] and on low-carbon benchmarks and positive carbon impact benchmarks.[52]

Conclusion: The EU's budget from 2014–2020 foresees that 20% of expenditure should be climate-related. The Commission has proposed to increase this to 25% for the budget 2021–2027. The Commission is actively promoting sustainable private finance as a tool to shape medium- and long-term investment in the low-carbon economy.

Conclusion

This chapter dealt with only a few policy areas of Europe's efforts to mainstream climate change into other policies and the budget of the European Union. Environmental policies to address the ozone hole have been highlighted, as have methane emissions from a variety of economic sectors, adaptation, EU spending programmes and the development of sustainable finance.

Over time, climate policy is likely to become more and more multifaceted. Gradually climate change will become a substantial part of the EU's standard policies such as on agriculture and trade, along with what has been achieved so far on energy, transport, research, industry and several other areas. Broad transformational change is afoot but still has a long way to go.

Notes

1 Regulation (EC) No 842/2006 of the European Parliament and of the Council on certain fluorinated greenhouse gases; OJ L 161, 14.6.2006, pp. 1–11: http://eur-lex. europa.eu/legal-content/EN/TXT/PDF/?uri=CELEX:32006R0842&from=EN.

2 Directive 2006/40/EC of the European Parliament and of the Council relating to emissions from air-conditioning systems in motor vehicles and amending Council Directive 70/156/EEC; OJ L 161, 14.6.2006, pp. 12–18. http://eur-lex.europa.eu/legal-content/EN/TXT/PDF/?uri=CELEX:32006L0040&from=EN.

3 Directive 2000/53/EC of the European Parliament and of the Council on end-of life vehicles; OJ L 269, 21.10.2000, pp. 34–43. http://eur-lex.europa.eu/resource.html?uri=cellar:02fa83cf-bf28-4afc-8f9f-eb201bd61813.0005.02/DOC_1&format=PDF.

4 European Electronic Component Manufacturers Association website 2019: https://www.eusemiconductors.eu/index.php?option=com_content&view=article&id=77&Itemid=166.

5 Regulation (EU) No 517/2014 of the European Parliament and of the Council of 16 April 2014 on fluorinated greenhouse gases and repealing Regulation (EC) No 842/2006; OJ L 150, 20.5.2014, pp. 195–230. http://eur-lex.europa.eu/legal-content/EN/TXT/PDF/?uri=CELEX:32014R0517&from=EN.

6 European Environment Agency report number 21/2018: *Fluorinated greenhouse gases 2018* (2018). https://www.eea.europa.eu/publications/fluorinated-greenhouse-gases-2018/.

7 European Commission: *Report from the Commission on barriers posed by codes, standards and legislation to using climate-friendly technologies in the refrigeration, air conditioning, heat pumps and foam sectors,* reference COM/2016/0749 final of 30.11.2016. https://eur-lex.europa.eu/legal-content/EN/TXT/?uri=CELEX:52016DC0749.

8 European Commission: *Report from the Commission on availability of training for service personnel regarding the safe handling of climate-friendly technologies replacing or reducing the use of fluorinated greenhouse gases,* reference COM/2016/0748 final of 30.11.2016. https://eur-lex.europa.eu/legal-content/EN/TXT/?uri=CELEX:52016DC0748.

9 European Commission: *Report from the Commission assessing the quota allocation method in accordance with Regulation (EU) No 517/2014,* reference COM(2017) 377 final of 13.07.2017 (see link on page: https://ec.europa.eu/clima/policies/f-gas/legislation_en#tab-0-2).

10 European Commission: *Report from the Commission assessing the 2022 requirement to avoid highly global warming Hydrofluorocarbons in some commercial refrigeration systems,* reference C(2017)5230 final of 04.08.2017 (see link on page: https://ec.europa.eu/clima/policies/f-gas/legislation_en#tab-0-2).

11 There can also be venting of methane, whether intentional (for maintenance or operational reasons) or unintentional.

12 Council Directive 1999/31/EC on the landfill of waste; OJ L 182, 16.7.1999, pp. 1–19: http://eur-lex.europa.eu/legal-content/EN/TXT/PDF/?uri=CELEX:31999L0031&from=en.

13 Directive (EU) 2016/2284 of the European Parliament and of the Council of 14 December 2016 on the reduction of national emissions of certain atmospheric pollutants, amending Directive 2003/35/EC and repealing Directive 2001/81/EC; OJ L 344, 17.12.2016, pp. 1–31.

14 Report from the Commission to the European Parliament, The Council, the European Economic and Social Committee and the Committee of Regions: The First Clean Air Outlook COM(2018)446 final of 07/06/2018.

15 Cofala, J., Klimont, Z., of the International Institute for Applied Systems Analysis (IIASA): *Emissions from households and other small combustion sources and their reduction potential,* TSAP Report #5 Version 1.0, June 2012. http://pure.iiasa.ac.at/id/eprint/10160/1/XO-12-015.pdf.

16 See http://europa.eu/rapid/press-release_MEMO-16-4372_en.htm.

17 Mauritsen, T., and Princus, R. (2017). *Committed Warming Inferred from Observation.* Nature Climate Change. Volume 7, September 2017 pp. 652-656. https://www.nature.com/articles/nclimate3357.pdf.

18 IPCC (2014). *Climate Change 2014: Synthesis Report. Contribution of Working Groups I, II and III to the Fifth Assessment Report of the Intergovernmental Panel on Climate Change.* Geneva: Intergovernmental Panel on Climate Change.

19 IPCC (2012). *Summary for Policymakers: Managing the Risks of Extreme Events and Disasters to Advance Climate Change Adaptation.* Cambridge: Cambridge University Press.

20 Article 7(1) of the Paris Agreement.

21 European Environment Agency (2017). *Climate Change, Impacts and Vulnerability in Europe 2016: An Indicator-Based Report.* Copenhagen: European Environment Agency.

22 EEA Report No 15/2017, "Climate Change Adaptation and Disaster Risk Reduction in Europe," updated in 2018 as part of the EEA indicator on "Impacts of extreme weather and climate related events in EEA member countries."

23 European Commission (2013). Communication from the Commission to the European Parliament, the Council, the European Economic and Social Committee and the Committee of the Regions, An EU Strategy on adaptation to climate change (COM(2013)216 final).

24 European Commission: webpage *LIFE Climate Action*: https://ec.europa.eu/clima/policies/budget/life_en.

25 European Commission report on the Implementation of the EU Strategy on adaptation to climate change (COM(2018)738 of 12.11.2018).

26 The Global Covenant of Mayors for Climate and Energy formally brings together the Covenant of Mayors and the Compact of Mayors, the world's two primary initiatives of cities and local governments to advance their transition to a low emission and climate resilient economy and to demonstrate their global impact. It incorporates, under a single umbrella, the commitments of individual cities and local governments put forth either through the Compact of Mayors or under regional/national Covenants, operating under principles and methods that best suit each region.

27 European Climate Adaptation Platform "Climate-ADAPT" is a partnership between the European Commission and the European Environment Agency. Details of the Platform can be found at: https://climate-adapt.eea.europa.eu/.

28 CEN-CENELEC (2016): CEN-CENELEC Guide 32: Guide for addressing climate change adaptation in standards.

29 European Investment Bank: *Integrating Climate Change Information and Adaptation in Project Development: Emerging Experience from Practitioners*, by the European Financing Institutions Working Group on Adaptation to Climate Change (version 1.0, May 2016). https://www.eib.org/attachments/press/integrating-climate-change-adaptation-in-project-development.pdf.

30 European Commission (2016): Climate Change and Major Projects: Outline of the climate change related requirements and guidance for major projects in the 2014–2020 programming period.

31 See Section 4 of Communication from the Commission to the European Parliament, the Council, the European Economic and Social Committee and the Committee of the Regions – A Roadmap for moving to a competitive low carbon economy in 2050 (COM/2011/0112 final of 08/03/2011).

32 European Commission: Joint Research Centre - Science for Policy Report (2018) "EU coal regions: opportunities and challenges ahead". https://ec.europa.eu/jrc/en/publication/eur-scientific-and-technical-research-reports/eu-coal-regions-opportunities-and-challenges-ahead.

33 European Commission: *Future Financing of the EU: Final report and recommendations of the High Level Group on Own Resources*" (December 2016). https://ec.europa.eu/budget/mff/hlgor/library/reports-communication/hlgor-report_20170104.pdf.

34 18.8% of EU budget for 2014–2020 is the estimated climate-related expenditure according to the European Commission's consolidated updated information on the 2014–2020 programming period (Source: Draft budget 2018, Statement of Estimates, page 101 onwards).

35 European Commission Communication "Mid-term review/revision of the multiannual financial framework 2014–2020: An EU budget focused on results COM" (2016)603 final of 14.09.2016 and accompanying Staff Working Document SWD (2016)299 final.

36 European Court of Auditors: Special Report Number 31/2016: *Spending at least one euro in every five from the EU Budget on climate action: ambitious work underway, but serious risk of falling short* (2016) EU Publications Office, Luxembourg. https://www.eca.europa.eu/en/Pages/DocItem.aspx?did=39853.

37 COM(2018)321.

38 COM(2018)392.

39 COM(2018)372.

40 COM(2018)435.

41 According to the 2018 report on World Energy Investments by the International Energy Agency "World Energy Investment 2018," China has become the world's largest destination of investment in the energy sector, standing for one-fifth of global energy investment in 2017.

42 Proposal for a Regulation establishing Horizon Europe – the Framework Programme for Research and Innovation, laying down its rules for participation and dissemination COM/2018/435 final.

43 Proposal for a Regulation on the European Regional Development Fund and on the Cohesion Fund COM/2018/372 final – 2018/0197 (COD).

44 Proposal for a Regulation of the European Parliament and of the Council establishing the InvestEU Programme COM/2018/439 final – 2018/0229 (COD).

45 Joint Research Centre (2018). "EU Coal Regions: Opportunities and Challenges Ahead".

46 Proposal for a Regulation on the financing, management and monitoring of the common agricultural policy and repealing Regulation (EU) No 1306/2013, COM/2018/393 final – 2018/0217 (COD).

47 Communication from the Commission to the European Parliament, the Council, the European Central Bank, the European Economic and Social Committee and the Committee of the Regions – Capital Markets Union – Accelerating Reform (COM/2016/0601 final of 14.9.2016).

48 European Commission website gives further details of this High-Level Expert Group, (established in 2016) including its membership and activity reports: https://ec.europa.eu/transparency/regexpert/index.cfm?do=groupDetail.group Detail&groupID=3485&NewSearch=1&NewSearch=1&Lang=EN.

49 Communication from the European Commission to the European Parliament, the Council, the European Central Bank, the European Economic and Social Committee and the Committee of the Regions – Action Plan: Financing Sustainable Growth (COM/2018/097 final of 8.3.2018).

50 COM(2018)353.

51 COM(2018)354.

52 COM(2018)355.

10

TEN PERSONAL REFLECTIONS ON THE DIFFICULT JOURNEY TOWARDS CLIMATE NEUTRALITY

Jos Delbeke

Introduction

This concluding chapter offers some more personal reflections on the challenges after 2030 and, more specifically, with the goal of climate-neutrality in mind. It is not intended to be exhaustive. The thoughts originate from personal involvement in European climate policy making over the last two and a half decades. They reflect a strong belief that much more ambitious policies are urgently required. The science has convincingly shown that the dramatic and most costly impacts of climate change can still be avoided if ambitious action is taken quickly. While much greater political will is needed, the EU has already demonstrated that ambitious policies and measures can be adopted even within the EU's complex decision-making context. The following pages suggest ideas for continuing this journey in the coming years.

This chapter is not referenced as previous chapters have been but assumes that the previous chapters have been read. Many of the facts and acronyms referred to have been explained earlier in the book, and so this chapter is deliberately written to be more discursive and less explanatory.

10.1 Societal megatrends and the climate challenge

Societal megatrends represent the context for the formulation of a comprehensive policy vision on how to reach climate-neutrality within a perspective of 2050. Some of these megatrends will make it easier to reach the goals of the Paris Agreement, but others will make it much more challenging.

The most worrying megatrend is undoubtedly the one related to demography. The global population continues to increase, although at a slower pace, and is heading towards 10 billion by the middle of this century. While in recent decades a large proportion of the world's population has been lifted out of poverty, billions now aspire to enjoy standards of living comparable to those in the richer parts of the world. This will unavoidably lead to a significant upward pressure on greenhouse gas emissions, whereas the climate science implies that these must be reduced to an annual average below two tonnes of CO_2eq per person. Given that per capita emissions in Europe are today in the region of seven tonnes per person and in the US 15 tonnes, there is clearly a huge challenge if the goals of the Paris Agreement are to be met.

The EU's share of the global population will fall to below 5% and so will its share of greenhouse gas emissions. That means the world cannot depend on Europe's efforts to reduce its emissions, important though they may be. Europe's contribution to global solutions will rather be in terms of climate-related policy learning, innovation, technological development and deployment, as well as its contribution to climate finance. At the same time as its share of the population is declining, it is likely that migration towards Europe will increase because of climate change. In particular, the increasing severity of droughts and desertification in regions such as Africa will create harsher living conditions for the people living there, who will look for other, better places to live.

Another megatrend is the emergence of new digital technologies, including artificial intelligence. The world is an increasingly connected place and digitalisation is radically changing the ways societies organise themselves. While there are many opportunities for improvement, new global connectivity coupled with demographic pressures are contributing to more unstable political environments. Migration is contributing to the emergence of more nationalistic and populist political tendencies. Instead of developing policies – including climate policies – that would reduce these pressures, this has paradoxically led to the opposite, where countries turn inwards, looking to stem immigration, question the value of overseas development aid and promote more nationalistic economic policies.

On top of this, there is growing geopolitical tension generally, to which climate change is adding a new dimension. In addition to the pressures to the east and south of Europe, there is a new one looming in the north. The disappearance of the Arctic summer ice makes new economic exploitation and navigation routes possible, in an area where territorial disputes have literally been frozen until now. The rapid warming of the Arctic is changing this centuries-old reality and is leading to territorial claims and tensions that are likely to exacerbate in the coming decades.

These megatrends are in addition to the one resulting from the fossil fuel-based economic model that underlies the rising concentration of greenhouse gases in the atmosphere. This climate megatrend calls for urgent policy responses that will need to be sustained for decades. The Paris Agreement of 2015 created a fresh basis for globally coordinated efforts by all countries, whether highly developed and post-industrial, rapidly industrialising and emerging or struggling with poverty and the effects of climate change. The universal agreement on the ambitious goal of limiting the global temperature increase to no more than 2°C was accompanied by the reconfirmation that the fight against climate change is a "common but differentiated responsibility," and that all countries have to contribute according to their "respective capabilities." Usefully, the Paris Agreement has moved beyond the outdated dichotomy between "developed" and "developing" countries, as these are over-simplistic categorisations that no longer correspond to the realities of the 21st century.

Conclusion: The megatrends related to demography, technology and geopolitics will all shape policies leading towards climate neutrality. They offer both challenges and opportunities. Crucial, however, is that all countries act urgently to limit or reduce greenhouse gas emissions in differentiated ways that reflect their capabilities but all combining their efforts towards a global transition away from fossil fuels towards more sustainable solutions.

10.2 The urgent need for the implementation of policy plans (NDCs)

The IPCC has shown the extent to which the climate change we are currently experiencing is due to the greenhouse gas emissions originating from the developed countries that industrialised earlier on the basis of fossil fuels such as coal, oil and gas. However, the science is also clear that over the coming decades the newly emerging economies will be the main source of greenhouse gas emissions. Within the emissions budget that a maximum warming of 2°C would allow, at least two-thirds has already been consumed. In recognition of this fact, developed countries have generally accepted that they must reduce their emissions much earlier and more substantially compared to the emerging economies. At the same time, it is clear that emerging economies must also start reducing their emissions very soon; otherwise, the 2°C target will never be within reach.

This leads to the observation that the implementation of the plans submitted as part of the ratification process to the Paris Agreement, the so-called "Nationally Determined Contributions," should now be at the core of future discussions. The European experience shows that implementation is hard work. A raft of policies and measures need to be introduced and implemented in order to have an effect on emissions, and these necessarily have an impact on the daily lives of companies and citizens. It is imperative, however, that promises are fulfilled. This also applies for countries showing reluctance to engage, such as the US. Climate policies and measures have to make a difference, which means forcing change to happen faster than would have happened anyway.

A common drive towards the implementation of the Paris Agreement will increasingly allow for mutual learning. The EU experience to date may be a useful source of learning for other UNFCCC Parties. The EU has already shown that collaboration across national borders can reduce short-term economic costs. Equally, cross-border collaboration can facilitate the transition process as it allows for dealing with distributional issues, in particular for regions that have to make more fundamental transitions. Making a "just transition" away from a coal-based energy sector, for example, needs funding mechanisms that others might contribute towards but which is to everyone's benefit in terms of climate change.

In the coming years, before the ambition review planned by the Paris Agreement for 2023, policy implementation should receive stronger attention by all stakeholders, much more than discussions about tightening existing targets. It is too often assumed that setting ambitious targets is a guarantee for action on the ground. On the contrary, meaningful ambitious targets are ones that build on practical implementation experience. The new transparency provisions of the Paris Agreement, as agreed in Katowice at COP24, are a most useful new tool in this respect. Transparency will allow comparisons to be made on progress, or the lack of it, and lessons can be learned and shared. Europe should expect to learn increasingly from others, not least about the new low-carbon technologies that are being developed and deployed across the globe.

The EU should continue to focus a considerable share of the climate finance it provides to assist third countries in the urgent task of policy design and implementation. This task is of capital importance, and policy attention globally should now move in this direction. In addition, the UNFCCC

Secretariat should increasingly focus on following up on implementation more than on annual negotiations. It should be considered whether to reduce the annual frequency of expensive Conferences of the Parties (COPs), as questions are increasingly asked about the necessity of this frequency. At a minimum, it could be envisaged to have alternating "implementation" and "negotiation" COPs.

Conclusion: The absolute priority for the coming years is to see the full implementation by all countries of the policy plans they submitted as part of the ratification of the Paris Agreement. A review of the targets in 2023 is only meaningful if based on the experience – and results – of practical policy implementation.

10.3 The need for a carbon price combined with local policies

The Paris Agreement delivered a global decision on the climate ambition level but left the more challenging task of developing policies to the Parties. It is unfortunate that it was not possible to agree at least on a limited number of strategic common policies.

Based on the European experience, the principle of putting a price on carbon would have been a strong candidate for such a common policy. Carbon pricing offers considerable flexibility to economic operators and hence is a system that realises emission reductions at lower costs. This is not denying the fact that many more policies and measures are required to deal with specific market barriers, but a price on carbon is basic and essential, as it sends a clear signal to all individuals and companies that greenhouse gas emissions no longer can be released into the atmosphere without any cost. The atmosphere is not a convenient, free dumping ground for waste gases. Moreover, an explicit price on carbon serves as a useful benchmark against which other policies can be measured. What is important is that all the tools used by governments should be coherent and pull in the same direction.

There are several ways to put an explicit price on carbon. Most countries that have introduced carbon pricing have established "cap-and-trade" schemes. Economists tend to prefer carbon taxes, although real-life policy experience shows that introducing taxes is even more difficult politically than introducing "cap-and-trade" systems. One key element that emerges from the experience of the EU's Emissions Trading System (EU ETS) is that it is of capital importance to address the distributional impacts in favour of lower-income Member States or regions, just as it is essential to address the concerns about competitiveness by emission-intensive and trade-exposed sectors. Addressing these elements in the EU helped overcome political or economic reticence and helped to establish a common carbon price for all companies covered by the system, whichever of the 31 participating countries companies are based in. The EU ETS covers only the emissions from large fixed industrial installations, but over time, more sectors such as road transport could be progressively included.

Today, several countries and regions that have established carbon-pricing systems are talking to each other about possible cooperation on a bilateral or regional basis. Informal talks have been held between the EU, Canada, California, China and New Zealand, and it is already clear that preserving industry's competitiveness is a concern for all. Many economists, such as the 2019 Nobel Prize Winner for Economics William D. Nordhaus, advocate for the formation of "carbon clubs" where national systems are linked to one another. The EU and Switzerland have decided to formally link their emissions trading systems. The more countries join carbon-pricing systems, the less need there will be for policies to address concerns about competitiveness or "carbon leakage." Today, some 52 national, regional and local carbon pricing systems exist, covering some 20% of global emissions. The latter percentage will significantly increase once China has rolled out its nationwide system as of 2020. These running systems could, over time, become the core of a kind of "carbon club" that would be open for others to join.

Cooperation between cap-and-trade systems is clearly the most promising route for establishing an international carbon market. As part of the discussions around Article 6 of the Paris Agreement, a fresh debate is taking place on how to complement these cap-and-trade systems with offset schemes. A strong argument in favour of international offsets is that they offer an incentive for emission reductions that are cheaper but also faster to achieve. Offsets systems, however, will always require solid proof that they are environmentally sound, managed in a transparent manner and are subject to strong compliance checks.

The Kyoto Protocol created international offset mechanisms, such as the Clean Development Mechanism (CDM) and the Joint Implementation (JI) projects. These offset mechanisms took off in a promising way and flourished as the EU opened its Emissions Trading System to these credits. As already explained in Chapter 4, the experience was a disappointing one, as the CDM Executive Board failed to impose a sufficiently strong governance system. Too many credits of doubtful environmental credibility were generated. The

EU decided to stop the influx of these credits as they undermined its carbon market. As a result, the price of credits collapsed to less than €1 a tonne of CO_2 equivalent and is likely to stay there as long as the question of robust governance remains unresolved.

In the meantime, international civil aviation may be the first global economic sector to create an offset regime, known as CORSIA, but to date, neither ICAO nor the UNFCCC has been able to adequately answer the fundamental questions relating to transparency and governance that underpins environmental integrity. It turns out, rather unsurprisingly, that the UN is not the right vehicle for organising markets. The best thing the UN bodies can contribute to, however, is creating the conditions for reliable monitoring, reporting and verification of emissions and ensuring that emissions are accounted for properly, avoiding any risk of double-counting. The mechanisms to ensure this between the UNFCCC and ICAO accounting systems are, unfortunately, far from robust enough. Quality control may well have to be delegated to trusted bodies that might develop "quality ratings" similar to those used in financial markets.

Moreover, an offset regime cannot escape pertinent questions about distributive impacts. The Paris Agreement requires comprehensive policy plans (NDCs) by all Parties. These plans relate to emission reductions achieved within the territorial limits of the country. It should therefore be clearly indicated which reductions give rise to offsets that are sold to emitters outside the national jurisdiction and so should be deducted from the reductions claimed under the national plan to avoid double-counting of emissions reductions (as reductions will also be claimed by the purchaser of the credits). National governments and stakeholders will be increasingly insistent that cheap abatement possibilities should not be sold off for the benefit of economic operators in other jurisdictions, leaving more expensive abatement options to be paid for by national economic actors or national governments. Rather than see an outflow of cheap credits, countries will in all probability wish to facilitate their own long-term transition to a low-carbon economy, while also fulfilling their international commitments.

Conclusion: Putting a price on carbon is fundamental. Regional cooperation brings the prospect of global carbon pricing within reach. Carbon offsets can be an additional and temporary solution, provided a solid governance system can be established. However, governments should be cautious towards allowing the benefits of cheap abatement options go to actors outside their jurisdiction in the form of offset credits, leaving more expensive options to be paid for domestically.

10.4 Towards a complete decarbonisation of the energy sector

The EU is accelerating its annual emissions reduction from approximately 0.5% in the 1990s to about 2% in the run-up to 2030. The energy sector has been at the heart of this effort, as it is in this sector in which the bulk of the fossil fuels are used. Efforts have therefore been concentrated on improving its efficiency and producing energy in less carbon-intensive ways.

The relatively low-carbon content of energy production in the EU, in particular in comparison to other highly industrialised nations, is linked to the build-up of nuclear capacity in Europe since the 1970s. However, the popularity of this source of energy production went down spectacularly after the major accidents at Three Mile Island (US), Chernobyl (Ukraine) and Fukushima (Japan). Today, more than 150 nuclear installations are still in operation in Europe. These major accidents, combined with the unsolved radioactive waste problem, led to a perception that nuclear power is an energy dream that never fully fulfilled its promises. These problems, as well as the high cost of new nuclear power generation, creates a situation today in which Europe is unlikely to opt for nuclear power generation as a major way to reduce greenhouse gases in the decades to come.

A major challenge for Europe is to strive for sufficient capacity in other new sources of low-carbon energy. The spectacular development of renewable energy is without a doubt one of the most important peacetime policy-driven changes to the energy landscape of the last decades and maybe even of the last century. Given the substantive cost reductions of sun and wind energy in particular, policy intervention is now shifting from subsidies to public regulation and the question on how to cope with their variability.

The drive towards reduced energy use combined with new low-carbon energy production will intensify in the future. Saving energy is a rational choice for the EU as it contributes almost €1 billion every day to energy imports. The drive towards more renewable energy is likely to be reinforced through spectacular developments and cost reductions in storage and battery technology and through the considerable potential of digitalisation. For Europe, therefore, the first significant step towards climate neutrality is, without a doubt, climate-neutral energy production.

Nevertheless, a major question about the other 68% non-renewable electricity production in 2030 remains squarely on the table. Given the unlikely increase of new nuclear capacity, an unavoidable choice is the technology known as "Carbon Capture and Storage." The use of lower-carbon fossil fuels such as natural gas could continue in the future, provided the carbon emissions would be captured and then stored in the underground. This technology is well known but not tested on a sufficiently large commercial scale yet; this should become a high priority. Moreover, it is expensive and consumes energy, and some continue to raise questions about the safe storage of carbon dioxide underground. On the other hand, gas fired power generation is flexible and allows for a good complement to the variable production of electricity through solar and wind.

The rest of the world is unlikely to follow the EU's energy experience, or at least will do so only partially. One can rather expect a huge interest in the new emerging economies in energy conservation and renewable energy technologies, and they may well concentrate their efforts in developing these options. However, their rapidly rising energy demand is likely to be satisfied by also adding a fair amount of nuclear capacity to the energy mix, notwithstanding the higher costs and the safety risks associated with nuclear technology. This will generate new questions about proliferation and waste handling.

Conclusion: By 2050, a climate-neutral energy system is technologically possible at the global level. Both the private and the public sectors should double their efforts. More renewables and the commercial viability of carbon capture and storage will be key.

10.5 Low-carbon transport is urgently required

Transport has not contributed sufficiently to reducing the EU's emissions so far. This must change radically if the EU is serious about meeting its 2030 targets, even more so if the goal is to attain carbon-neutrality by 2050. Economists would say that the real problem with transport is that the income elasticity of demand is higher than one. That means that the demand for transport services increases more than proportionally as people become richer. That does not bode well for the rapidly emerging economies, where transport demand has been rising rapidly and huge investments in infrastructure have barely been sufficient to avoid severe traffic congestion and air quality problems.

The EU has been adopting ambitious CO_2 standards for cars and lorries, and many new technologies are now on the brink of uptake by the market. A variety of new technologies can be expected in the coming years, including on hydrogen and low-carbon fuels. However, the shift towards the electrification of transport is likely to be the most profound transformation over the coming decade, and this fits well with a power sector that is rapidly decarbonising.

For both passengers and freight transport, much more additional action at national and local levels is needed to ensure that transport systems fit with the growing public expectation of liveable, sustainable cities with substantially improved air quality. Public infrastructure is often in need of major overhauls, including the provision of walking and biking lanes or with respect to electric charging infrastructure. A variety of local traffic management measures will also have to be considered, adapted to the local circumstances.

Transport needs to be fluid and efficient. More generalised road pricing policies are becoming unavoidable, however difficult it may be to win public acceptance of these. The political discourse around road pricing has to move away from one that centres on fiscal charges towards a "pay-for-service" approach whereby paying the charge gives access to a faster and more efficient journey. For freight transport, the "Eurovignette" system has been gradually developed for heavy goods vehicles and over time needs to serve as the foundation for a more generalised system of road pricing that Member States can decide to also apply to passenger cars.

New technologies are always somewhat more expensive and fiscal policies can be geared towards speeding up the low-carbon transition. Europe is also endowed with a magnificently dense network of railway lines that need to be exploited more efficiently. Rail transport has tremendous potential as a supplier of low-carbon transport services but is disadvantaged by management systems that belong to another age. These issues are being addressed too slowly, giving rise to perverse relative prices for users between different transport modes, in particular the relative costs between rail and aviation services for distances below 1000 kilometres.

Given the rapidly rising income levels in emerging economies and growing urbanisation, very large investments in transport facilities and infrastructure will be built in the coming decades. While many countries do not have the dense railway network that Europe has, other options than building highways or expanding domestic aviation have to be considered in the light of rapidly growing demand for mobility. There are alternatives, as the development of rail infrastructure in China demonstrates.

Conclusion: The road transport sector is ready to introduce and scale up new low-carbon technologies. Transport policies still need a major overhaul at all levels of governance: improved traffic management, a review of national taxation regimes, addressing major deficiencies in local and city infrastructure and modernising the rail sector. In a world of rising mobility demand, we cannot afford for transport to wipe out hard-won emission reductions made in other sectors.

10.6 Emissions from aviation and maritime sectors grow out of control

The real nightmare for transport is the seemingly unstoppable growth of emissions from the aviation and maritime sectors. The actions of ICAO and IMO, despite intensive pressure by the EU, are too piecemeal and misaligned with the environmental objectives of UNFCCC and the Paris Agreement.

The introduction of the CORSIA system to offset the growth of international aviation's emissions from 2020 is expressly voluntary until 2027, and the governance system underpinning its environmental integrity is extremely weak. The prospect of energy efficiency gains of aircraft, improvements in air traffic management and sustainable aviation fuels actually reducing emissions from aviation, in the light of strongly increasing demand, are – to put it politely – remote. More will have to be done.

International shipping's CO_2 emissions may have peaked in 2008, but it is difficult to see international shipping halving its total greenhouse gas emissions by 2050, as is the IMO's declared aim. The IMO is working on a long list of possible instruments and candidate measures, but there is no agreement yet on which of these might be implemented, nor any certainty on how stringent such measures will be.

One must expect, therefore, that countries will listen to growing public concern about the sustainability of aviation and shipping and put in place further policy measures, irrespective of whatever global approaches are agreed in ICAO or the IMO. One example for aviation is passenger levies, such as they are already applied in a number of European countries. Why aviation is not subjected to Value Added Tax on the sale of tickets is surprising, to say the least. Air travel may be democratising, but most air passenger transport is used by wealthier segments of the population.

Another option is the legislation on the EU ETS that already foresees the inclusion of both incoming and departing flights to and from the EU as of 2024. The EU ETS could also be applied to international shipping if sufficient action is not taken in the IMO. Indeed, the EU is committed to assess the adequacy of actions taken in ICAO and the IMO in the context of future amendments to the EU ETS. Unless meaningful action is undertaken at the international level, which looks unlikely, then the EU will have no choice but to apply additional polices and measures to these international modes of transport.

A major responsibility for doing more rests on the shoulders of ICAO and the IMO, United Nations agencies whose legitimacy is not just to defend the sectoral interests of the industries they represent, but who ought to defend the public interest. This will require them to reinvent themselves in the light of the environmental challenges faced. Motivated millennials and other segments of the population are beginning to see aviation and shipping as contributing disproportionately to a problem that the younger generation will be left to suffer from – and pay for – the consequences. It should be no surprise that more people are seeking ways to avoid flying if possible. Cruise ships are increasingly seen as major sources of pollution and the epitome of unsustainable travel for amusement. If buying locally sourced goods is seen both as better for the environment and protecting local jobs, the momentum behind such initiatives could grow at the same time as more protectionist trade policies are advocated in some jurisdictions.

Conclusion: The aviation and maritime sectors do not pay for the external costs of their operations. On the contrary, they enjoy a most favourable fiscal treatment compared to other modes of transport. National and regional authorities will be under increasing pressure to remedy this situation in the absence of sufficient action by the international bodies such as ICAO and IMO.

10.7 The need for an enlightened industrial and trade policy

Another major long-term challenge is the low-carbon production of industrial commodities such as cement, steel and non-ferrous products, chemicals, pulp and paper and glass. The production of these commodities is very carbon-intensive, while these products are, and will continue to be, needed in huge quantities, for example to build the sustainable infrastructure and the cities of the future. Over the past years, significant results have been achieved as regards energy savings, but in view of reaching climate neutrality, completely new production processes will be required. For example, the blast furnace is a centuries-old process that emits very substantial amounts of carbon during steel production, even if gradual technological improvements have reduced the carbon emissions significantly.

This comprehensive transformation process needs an enlightened industrial policy. The EU ETS provides both an incentive to reduce emissions and a reward – in the form of comparative advantage – for those companies who emit less. As a result, technological efforts have already been stepped up significantly. Research indicates that the number of patents for low-carbon technologies has been increased significantly since the creation of the EU ETS. Innovation policy by both the public and private sector must ensure that the new generation of low-carbon technology leaves the laboratory sooner rather than later. The EU ETS Innovation Fund has been set up to facilitate this process and will have an envelope of more than EUR 10 billion to facilitate the transition towards the commercial uptake of promising new technologies.

The transformation of the industrial tissue will require considerable investments and hence will need a solid finance system geared up to the task. Many interesting ideas are being raised to promote green finance. Equally important may be the macroeconomic support for low-carbon investment, for example through making the Euro area's Stability and Growth Pact criteria more flexible for low-carbon investment. In addition, ways may be found to link cash lying in savings accounts to help finance productive low-carbon investments with more stable long-term returns.

This new approach to innovation in industrial sectors needs to be complemented by an enlightened trade policy. One can assume that new technology in its early phases will be costly and that the free allocations under the EU ETS and the Innovation Fund will only be able to compensate for part of this additional cost. However, industrial goods are intensively traded internationally. As a similar carbon constraint to the EU ETS, which is unlikely to be applicable to similar commodities produced in many other countries, it can be expected that discussions around trade corrections will intensify. Such discussions will need to be orderly and by far the best solution would be to organise these in the context of a reformed WTO. Alternatively and in the meantime, the EU could also consider the incorporation of climate policy implementation checks as part of its bilateral and regional trade agreements. So far, only recognition of the Paris Agreement is being explicitly mentioned, but there are no checks on whether the commitments made under the Paris Agreement are being implemented in practice. Such checks would be particularly relevant for more intensively traded goods and commodities.

These trade concerns would be much less acute in a world where all countries were moving towards the implementation of the Paris Agreement. Ideally, therefore, the implementation of the Paris Agreement will take place according to the so-called Nationally Determined Contributions. Businesses will then be making efforts that are more comparable as far as their implicit carbon costs are concerned. However, some countries will remain slow to act, or may even choose to "free-ride." An alternative could therefore be the creation of carbon clubs, as already mentioned.

One can also expect more competition on new low-carbon technologies. Asia has already taken a strong leadership position in sustainable technologies, such as batteries or high-speed trains. China has targeted several sectors in which it deliberately seeks to develop leadership to contribute to the greening of its own domestic economy and to position itself in fast growing segments of a new international market for low-carbon goods.

Conclusion: The low-carbon transformation requires a review of the research, innovation, finance and trade policies for intensively traded goods. A globally coordinated transformation, under the auspices of a reformed WTO for example, will allow for a much smoother implementation of the Paris Agreement.

10.8 The increasing importance of agriculture and forestry

As emissions from fossil fuels use reduce over time, attention will necessarily shift to other greenhouse gases, such as methane, or to more active ways of absorbing carbon from the atmosphere. This brings agriculture and forestry to the forefront of attention.

A major source of methane is meat production and in particular cattle and sheep rearing. Given the growing world population and in particular the growth of its middle classes, a switch in people's diets is happening. A global generalisation of the high levels of meat consumption of wealthier parts of the world is worrisome and is just not compatible with the temperature goals of the Paris Agreement.

Even if technology can help to some extent through improved farming methods, the major transformation will have to come from changing habits. The example of smoking has shown not only how important societal changes can be achieved but also how much time and effort this requires. In Europe, meat consumption has started to falter as a result of a new impetus to prefer vegetarian or vegan food. In other parts of the world, such as in Asia or Africa, the consumption of white meat is more established, and this habit should be maintained if possible. At the same time, more educated middle classes everywhere need to be made much more aware of the beneficial impact of eating less red meat.

The changing food consumption towards less meat should allow for less need of agricultural land. This could slow down and possibly halt the clearing of tropical forests, and existing land could be used for afforestation. Together with sustainable forestry, the carbon uptake could be much higher compared to what is currently the case. This may require new forms of remuneration;

in this context, the use of offsets is frequently mentioned. In addition, new technological developments can help to set up a solid governance system, as earth observation from space with better accuracy is becoming increasingly available.

In a longer-term perspective, the use of bio-energy with carbon capture and storage (BECCS) is a promising new technology. In view of reaching climate neutrality, it would allow for negative emissions and compensate for those emissions that are impossible to avoid. More practical experience of these technologies is urgently needed.

Conclusion: Increased policy attention on the agriculture and forestry sectors is indispensable to reach climate neutrality. Food habits will have to change in favour of consuming much less red meat. Carbon sinks must be enhanced through maintaining tropical forests, increased afforestation and sustainable forest management.

10.9 The critical importance of local authorities and citizens

An ambitious implementation of the Nationally Determined Contributions submitted under the Paris Agreement is not a task the public sector can deliver on its own. The magnitude of the task requires collaboration between multiple proactive actors in all segments of society, such as businesses, investors, bankers, consumers, citizens, transport users, cities and local authorities. The intention of making comprehensive policy plans is to ensure that all actors pull in the same direction, as well as to avoid short-term interests and disruption being used as an excuse for no action.

Cities and local authorities will have a particularly important role in policy making in the coming years, given the importance of the investment decisions they make and their influence. There is immense pressure on them to adapt local infrastructure to both reduce emissions, for example through energy efficient housing and transport, and to protect against the impacts of climate change. Increasingly, they will need to protect citizens against the negative effects of climate change such as storms or flash flooding. It is therefore at the city and local level that perhaps the most important action needs to take place. In such a situation, it will be very important to share experiences between municipalities and local authorities, to make sure that policies work together in a coherent manner. National and regional organisations can facilitate this knowledge sharing by careful monitoring and review of what works best.

Much more action can also be expected from the private sector. Important initiatives are under way as shareholders increasingly ask questions about the carbon content of the portfolio of activities of their companies, and they are already raising their voice in general assemblies. Pension and sovereign wealth funds are starting to redirect their long-term investment portfolios according to sustainability criteria in order to safeguard profitability over time. Bankers are under pressure from their clients to offer green savings products that support low-carbon activities. The first regulatory activities are under way in the European Union, but also in China, to provide greater information about the environmental impacts and potential liabilities of businesses, as well as to enhance the clarity and reliability of "green" investment products offered.

Consumers also become increasingly aware of the environmental impacts of the products they use and how these are minimised. This ranges from vegetables or flowers transported from faraway places, compared to products produced using shorter production chains closer to consumers. Public authorities have a particular responsibility to make transport more sustainable, whether they find themselves close to the consumer, such as municipal and local authorities, or far away such as ICAO and IMO when it comes to international aviation or maritime activities.

Fiscal provisions can be developed to create temporary favourable conditions for new promising technologies that are often more expensive in the early phases of their commercialisation, such as electric vehicles or heat pumps. Such incentives can help to switch a niche market into a mainstream activity and through this create scale effects that bring down unit costs. The example of solar panels and wind turbines is very encouraging in this regard. Public authorities also learned a useful lesson that such incentives can only be offered on a temporary basis and need to be monitored closely in order to avoid perverse effects. In many parts of Europe, incentives for renewable technology were maintained for too long and created a negative public reaction when the magnitude of the real costs became clear.

Another major new field for action by local authorities is the creation of infrastructure required to facilitate the recycling and reuse of consumer products. Encouraging circular economic activities would allow for a significant reduction in the energy and carbon intensity of industrial goods.

This requires the setup of an entirely new infrastructure for the collection and sorting of waste, the creation of new jobs and the making of new investments. These new activities need to be financed through levies consistent with the polluter-pays principle.

Conclusion: Local authorities and citizens, as well as investors and companies, play a critical role in joint low-carbon action in the sectors of housing, transport, private consumption and public services related to the circular economy.

10.10 The growing challenge of adaptation to climate change

Even if we are to limit the global average temperature increase to well below 2°C, this will unavoidably cause many effects of climate change, including some grave and dangerous effects. Today we have passed the 1°C mark, and the impacts of climate change have started becoming visible to everyone. It also becomes a painful reality for public authorities, mostly at the local level, who have to cope with more frequent flooding, forest fires, shortages of drinking water, severe storms and so on. A multitude of challenges falls on their shoulders.

The new local policies related to the adaptation to climate change, combined with the policies to bringing down greenhouse gas emissions, are profoundly changing the tasks of local authorities and each of these new tasks need to be tailored to the specific local circumstances. As they can learn a lot from each other's experiences, forums have been established where they can share and compare each other's experiences, such as through the Global Covenant of Mayors initiative.

National and regional authorities are already preparing for new challenges. Disaster risk reduction strategies are being developed at local, national and international levels. Infrastructure that is built to last several decades will increasingly need to be made to withstand extreme weather events and be made as "climate proof" as possible. It is hardly conceivable to build new rail links or highways in areas that may be prone to frequent flooding, for example. At the same time, the private sector, and in particular the insurance sector, is already increasingly confronted with new realities. Insurance is a service involving the transfer of financial risks against payment. Insurers and re-insurers are increasingly warning that climate change will cause more extreme weather events, and premiums will have to rise to reflect the increasing risks.

Conclusion: As climate change is already happening, public policies and infrastructures need to be "climate proofed." Adaptation to climate change will have to be tailored to specific local realities, but an overall increasing need for preparedness to counter climate-related risks is already inevitable. The less we do to reduce emissions of greenhouse gases, the more we will have to invest in adaptation.

Conclusion

The journey towards climate neutrality is going to be long and difficult. The IPCC offered a sobering message in its 2018 report on how limited the emissions window has become to keep the 1.5°C objective within reach. We know the magnitude of the task ahead and the hurdles to be overcome. The world has already passed the mark of 1°C global warming. If we are to succeed in avoiding dangerous climate change, with possible run-away climate change, a significant reinforcement of climate policies is urgently needed. It requires action by all countries, and the developed countries must be ready to deliver deep reductions. However, just as important is the extent to which emerging economies are able to limit and eventually reduce their emissions.

Even if we know that the challenge is immense and daunting, we have nevertheless already learned some useful lessons. It is important to focus on the low-cost potentials to reduce emissions. However, in parallel, we need to prepare the future through innovation and the deployment of new technologies. We need many technological breakthroughs, but even more so, we need a raft of social and institutional innovations. "Path dependency" is a brake on the speed of change in all aspects of life. Economies have prospered on energy from fossil fuels. Feeding an ever more populous planet is increasingly difficult. We have so many habits and ways of doing things left over from an era when the world was less populated and less industrialised, in which climate change was unknown. Our understanding of climate change, and what needs to be done, is gradually changing the nature of value creation in our economies. Change has already started. Fortunately, these will replace jobs in some sectors with other kinds of jobs and other skills.

At the same time, we have to widen the scope of action. So far, we have concentrated action on the energy, transport and industrial sectors, since their use of fossil fuels are the main sources of greenhouse gas emissions. Even if a lot more needs to be done in these sectors, we already know that many more sectors need to be involved, as the challenge is much wider and deeper. We have to extend efforts in the agriculture and land use sectors and look for carbon sinks in the soil and in trees. We also have to open up the promises of the "circular economy," encourage the recycling of carbon emissions by their incorporation into synthetic fuels, for example, and ultimately develop permanent greenhouse gas storage solutions. Once we have found ways on how to do this at a much lower cost, we can actively roll out the dimension of the so-called "negative emissions."

We also know now, much more than before, that we have to adapt to the very diverse manifestations of the impacts of climate change. We know how expensive this might become, as higher storm surge and flood protections need to be built, and new urban infrastructure will be needed in response to higher summertime temperatures and so on. While the richer parts of the world will be better able to afford this, a disproportional impact on the lower-income parts of the world needs to receive special attention. In particular, the way development assistance is organised and deployed will have to be revisited.

Implementation of the agreed upon Paris Agreement commitments will require a fundamental rethinking of public policies that have sometimes been set up for decades, if not centuries. We started this fundamental review of public policies incrementally, perhaps too timidly, but we have at least been able to learn some useful lessons in the meantime. There is no other alternative than to build further on this policy learning and to prepare for the next phases in a much more decisive and determined manner. In this respect, full attention needs to be paid to the social and geographic redistributional effects of these policies. All need to be carried along in finding solutions to one of the greatest challenges of our time.

INDEX

Note: Page numbers in *italics* indicate figures and in **bold** indicate tables on the corresponding pages.

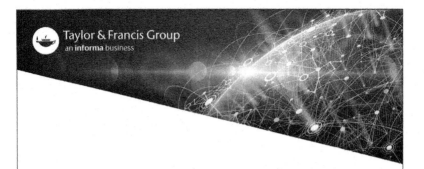